TOP OFFICE IV
顶级办公 IV

大型办公 创意办公 LOFT办公　DAM工作室 主编

华中科技大学出版社
http://www.hustp.com
中国·武汉

Preface 序言

Offices become the center of modern people's life, and the interior design of offices exercise an immediate influence on the staff's mental state and work efficiency. Traditional offices with single function are only work containers, modern offices, whereas, owning to their multiple functions, are medium of work and communication. With the development of society and technology, the internet, intelligent design and the idea of ecology will change the ways of life, work and study.

Ecologicalized offices are spaces where people and nature reach a perfect combination. An ecological environment is created so that people in offices can enjoy ample sunshine, fresh air and amazing scenery, which makes it possible for the staff to work with freedom and vigor.

Personalized offices embody the idea of humanized design. Nowadays, personalization and personal style are emphasized, and people want to voice characteristics with featured forms. This want in offices promotes the formation of new office culture which pays more attention to humanistic spirit.

Combining cooperate cultures, brands and cooperate images, we create representative offices of features by analyzing the category and orientation of different enterprises. This book includes various offices with different functions, such as listed groups, small enterprises and SOHO offices.

In this book, various kinds of projects are displayed and summed up to demonstrate ideas and techniques of design and decoration from different angles.

办公空间是现代人生活的中心，办公空间的室内设计直接关系到员工的工作心理状态和工作效率。传统的办公空间功能单一，是工作的"容器"，现代新型办公空间的功能多样化，成了人们工作和交往的"媒介"。随着社会的发展与科技的进步，网络、智能与生态的发展将再次改变人类的生活、工作与学习的方式。

生态化办公空间主张人与自然的完美结合，力求在办公区域营造出生态的环境，让使用者能够享受到充足的阳光，呼吸到新鲜的空气，观赏到迷人的景色，让每个人都能以洒脱的心情、旺盛的精力投入到工作中。

个性化办公空间是人性化设计思想的体现，个人及企业都在强调个性化和个人风格，渴望用具有特点的形态表达个性，而办公空间内新办公文化的衍生，也使人文精神进入了新的境界。

在设计不同类型、不同功能的办公空间时，如上市集团的办公空间，或创业初期及SOHO族的办公空间，要分析不同的企业类别及其公司定位，结合企业文化、品牌、企业形象等，再不断推敲与提炼，塑造出独具特色的代表性空间。

本书对大型办公空间、创意办公空间、LOFT办公空间三种类型的案例进行归纳与展示，从不同角度体现办公空间装饰与设计理念以及设计手法。

深圳市伊派室内设计有限公司设计总监　段文娟

Foreword 前言

Foreword 前言

办公空间是现代人生活的中心，办公空间的室内设计直接关系到员工的工作心理状态和工作效率。传统的办公空间功能单一，是工作的"容器"，现代新型办公空间的功能多样化，成了人们工作和交往的"媒介"。随着社会的发展与科技的进步，网络、智能与生态的发展将再次改变人类的生活、工作与学习的方式。

生态化办公空间主张人与自然的完美结合，力求在办公区域营造出生态的环境，让使用者能够享受到充足的阳光，呼吸到新鲜的空气，观赏到迷人的景色，让每个人都能以洒脱的心情、旺盛的精力投入到工作中。

个性化办公空间是人性化设计思想的体现，个人及企业都在强调个性化和个人风格，渴望用具有特点的形态表达个性，而办公空间内新办公文化的衍生，也使人文精神进入了新的境界。

在设计不同类型、不同功能的办公空间时，如上市集团的办公空间，或创业初期及 SOHO 族的办公空间，要分析不同的企业类别及其公司定位，结合企业文化、品牌、企业形象等，再不断推敲与提炼，塑造出独具特色的代表性空间。

本书对大型办公空间、创意办公空间、LOFT 办公空间三种类型的案例进行归纳与展示，从不同角度体现办公空间装饰与设计理念以及设计手法。

深圳市伊派室内设计有限公司设计总监　段文娟

Each and every design is customized, and the design of offices needs more. In designing offices, we consider corporate cultures, divides of functional requirements, working environment, and the most important thing, the relationship between people and spaces.

It is very important for the staff or people who come into the spaces to live in harmony with offices, since offices are places where people spend most time out of home, some people may spend more time in offices than at home.

Once receiving a design commission, we get to know the corporate culture, operation principle and functional needs at first, and next we design an office which is in harmony with the environment and humanity.

Many people asked me what is the most important part in a public space design, and my answer is, for a commercial space design, the most important thing is to build a space which is accord with the corporate function and idea with the most economical cost and best technique of aesthetic. Building a space is not stacking things up, and offices should be created with the simplest skills to present the essence of corporate culture.

Nowadays, armed with sophisticated ideas and skills, most designers are good at carrying out their requirements for aesthetic and functions on the base of environmental protection, energy conservation, ecology and humanity to perfectly interpret their understanding and perception of spaces.

Chinese culture has stepped onto the international stage, and many foreign designers having Chinese culture in mind have applied Chinese elements into designing spaces, which inspires Chinese designers to integrate their own culture into their design. What's more, Chinese designers have begun to pay attention to the relationship between the environment and the light-and-shadow and the spaces. They bring their perception of gardens into the interior, and make spaces where light and shadow, people and nature, environment and materials are in harmony, thus create working spaces which are better for thinking.

所有的设计都是量身定制的，办公空间的设计更是如此。在设计任何一个空间时，都要考虑到企业的企业文化、功能需求、工作氛围，以及人与空间的关系。

在办公空间设计中，我们尤其应该注重人与空间的关系，身处其中的人是否能与该空间和谐相处至关重要，因为一天之中的大部分时间员工都身处办公空间之内，甚至有时候比在家的时间还长。

当我们接到一个办公空间的设计委托案时，我们首先要做的是了解该公司的企业文化、经营理念，以及功能需求。然后在这些前提下，运用专业知识去感受该企业的整个氛围，再运用到空间之中，让整个空间与环境、人文和谐共融。

很多人问我，公共空间设计最重要的是哪一部分。我认为，商业空间设计最重要的是，如何用最低的造价，用最符合美学的手法打造出符合该企业功能理念的空间。空间打造并不是简单的堆砌，从某些角度来讲，应该是简练有力，干净利落，用最简单的手法来表达该企业的文化精粹。

现阶段，大部分设计师的设计理念和设计手法越来越成熟，也更善于在环保、节能、生态及人文的基础上把自己对美学及功能的需求细致化，从而更完美地诠释出他们对空间的理解和感悟。

当下，中国本土的文化已经渐渐走向国际舞台，很多国外设计师在关注东方文化的同时，已经在用他们独特的视角把中式元素运用于在各个空间之中。而本土设计师也慢慢地意识到了这一点，所以很多设计师在设计办公空间的时候也开始把中式元素融入空间文化里，开始注重环境与空间的关系，光影与空间的关系，开始将对建筑及园林的理解带进室内，从而让空间更流畅，达到光与影、人与自然、环境与材料运用的和谐，从而更完美地打造出更适合思考的工作空间。

CEX鸿文空间设计有限公司创始人 郑展鸿

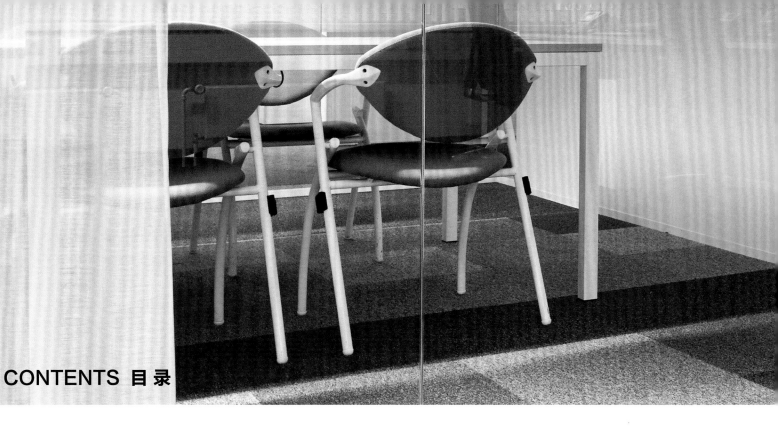

CONTENTS 目录

1 Large Office Space
大型办公

010	Genesis 商厦大堂及创意工作区	Genesis Commercial Building Lobby & Creative Working Space
018	致同国际办公室	Grant Thornton Office
026	大众汽车办公室	Volkswagen Office
032	朴本原美办公室	Beautiful Simplicity Office
038	HSB 办公室	HSB Office
050	Digital Agency 全球总部	Digital Agency HQ
056	NUAC 办公空间	NUAC Office
062	锦会商务办公中心	Shenzhen Jinhui Business Center
070	一起设计	Design Together
078	深圳阿基米联合办公中心	Shenzhen Archime Co-working Center
086	D3 交互式的环境	D3 Interactive Environment
092	Quadria Capital 新加坡办公室	Quadria Capital Singapore Office
100	印度 SAB Miller 办公室	SAB Miller India
112	Maxim Integrated 办公室	Corporate office for Maxim Integrated
118	日本某电子公司	Japanese E-commerce Company
124	easyCredit 总部	easyCredit HQ
138	BBDO 印度尼西亚办公室	BBDO Indonesia Office
148	T2 总部办公室	T2 Headquarters
154	Coop 集团办公室	Coop Office

2 Creative Office Space
创意办公

162	汉威士集团 / 阿诺国际传播波士顿总部	Havas/Arnold Worldwide Boston Headquarters
170	百利文仪杭州分公司办公室	Victory Hangzhou Branch Office
176	Masisa 办公室	Masisa Office

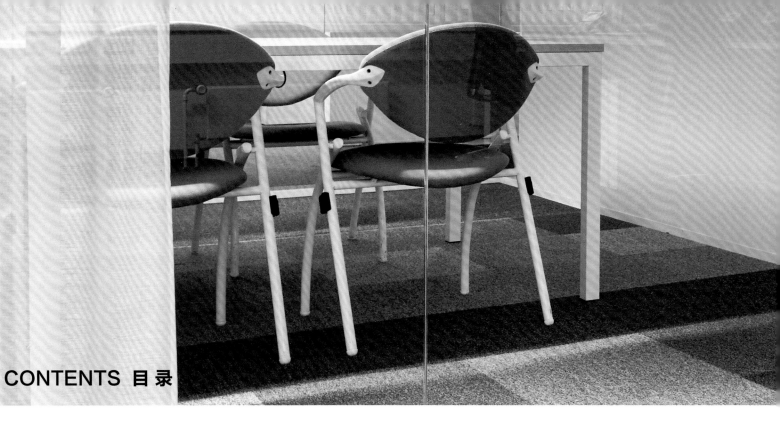

CONTENTS 目录

1 Large Office Space
大型办公

010	Genesis 商厦大堂及创意工作区	Genesis Commercial Building Lobby & Creative Working Space
018	致同国际办公室	Grant Thornton Office
026	大众汽车办公室	Volkswagen Office
032	朴本原美办公室	Beautiful Simplicity Office
038	HSB 办公室	HSB Office
050	Digital Agency 全球总部	Digital Agency HQ
056	NUAC 办公空间	NUAC Office
062	锦会商务办公中心	Shenzhen Jinhui Business Center
070	一起设计	Design Together
078	深圳阿基米联合办公中心	Shenzhen Archime Co-working Center
086	D3 交互式的环境	D3 Interactive Environment
092	Quadria Capital 新加坡办公室	Quadria Capital Singapore Office
100	印度 SAB Miller 办公室	SAB Miller India
112	Maxim Integrated 办公室	Corporate office for Maxim Integrated
118	日本某电子公司	Japanese E-commerce Company
124	easyCredit 总部	easyCredit HQ
138	BBDO 印度尼西亚办公室	BBDO Indonesia Office
148	T2 总部办公室	T2 Headquarters
154	Coop 集团办公室	Coop Office

2 Creative Office Space
创意办公

162	汉威士集团 / 阿诺国际传播波士顿总部	Havas/Arnold Worldwide Boston Headquarters
170	百利文仪杭州分公司办公室	Victory Hangzhou Branch Office
176	Masisa 办公室	Masisa Office

Each and every design is customized, and the design of offices needs more. In designing offices, we consider corporate cultures, divides of functional requirements, working environment, and the most important thing, the relationship between people and spaces.

It is very important for the staff or people who come into the spaces to live in harmony with offices, since offices are places where people spend most time out of home, some people may spend more time in offices than at home.

Once receiving a design commission, we get to know the corporate culture, operation principle and functional needs at first, and next we design an office which is in harmony with the environment and humanity.

Many people asked me what is the most important part in a public space design, and my answer is, for a commercial space design, the most important thing is to build a space which is accord with the corporate function and idea with the most economical cost and best technique of aesthetic. Building a space is not stacking things up, and offices should be created with the simplest skills to present the essence of corporate culture.

Nowadays, armed with sophisticated ideas and skills, most designers are good at carrying out their requirements for aesthetic and functions on the base of environmental protection, energy conservation, ecology and humanity to perfectly interpret their understanding and perception of spaces.

Chinese culture has stepped onto the international stage, and many foreign designers having Chinese culture in mind have applied Chinese elements into designing spaces, which inspires Chinese designers to integrate their own culture into their design. What's more, Chinese designers have begun to pay attention to the relationship between the environment and the light-and-shadow and the spaces. They bring their perception of gardens into the interior, and make spaces where light and shadow, people and nature, environment and materials are in harmony, thus create working spaces which are better for thinking.

所有的设计都是量身定制的，办公空间的设计更是如此。在设计任何一个空间时，都要考虑到企业的企业文化、功能需求、工作氛围，以及人与空间的关系。

在办公空间设计中，我们尤其应该注重人与空间的关系，身处其中的人是否能与该空间和谐相处至关重要，因为一天之中的大部分时间员工都身处办公空间之内，甚至有时候比在家的时间还长。

当我们接到一个办公空间的设计委托案时，我们首先要做的是了解该公司的企业文化、经营理念，以及功能需求。然后在这些前提下，运用专业知识去感受该企业的整个氛围，再运用到空间之中，让整个空间与环境、人文和谐共融。

很多人问我，公共空间设计最重要的是哪一部分。我认为，商业空间设计最重要的是，如何用最低的造价，用最符合美学的手法打造出符合该企业功能理念的空间。空间打造并不是简单的堆砌，从某些角度来讲，应该是简练有力，干净利落，用最简单的手法来表达该企业的文化精粹。

现阶段，大部分设计师的设计理念和设计手法越来越成熟，也更善于在环保、节能、生态及人文的基础上把自己对美学及功能的需求细致化，从而更完美地诠释出他们对空间的理解和感悟。

当下，中国本土的文化已经渐渐走向国际舞台，很多国外设计师在关注东方文化的同时，已经在用他们独特的视角把中式元素运用于在各个空间之中。而本土设计师也慢慢地意识到了这一点，所以很多设计师在设计办公空间的时候也开始把中式元素融入空间文化里，开始注重环境与空间的关系，光影与空间的关系，开始将对建筑及园林的理解带进室内，从而让空间更流畅，达到光与影、人与自然、环境与材料运用的和谐，从而更完美地打造出更适合思考的工作空间。

CEX鸿文空间设计有限公司创始人 郑展鸿

184	鸿文空间	Hongwen Space
194	Infocomm 投资办公室	Infocomm Investments (BASH) Office
200	美的·林城时代办公室样板房	Midea Forrest City Times Model Office
206	Payguru 办公室	Payguru Office
216	埃因霍温某办公室	Municipality Eindhoven Office
226	维斯林室内建筑设计有限公司办公室	Pplusp Designers Office
234	意大利电信公司	Italia Telecom Office
240	红牛办公室	Red Bull Office
250	Optimedia 媒体机构办公室	Optimedia Media Agency Office
256	迪欧家具办公室	Dious Furniture Office
262	百利文仪（中国）家具有限公司	Victory Furniture Group
268	中梁 V 城市办公样板房	Zhongliang V City Model Office
276	创域办公室	Creative Space Office

3　Loft Office Space / LOFT 办公

286	中梁 C 户型办公样板房	Zhongliang Model Office C
292	成都万科钻石广场 LOFT	Chengdu Vanke Diamond Plaza Loft
298	珠江科技数码城 LOFT 公寓	Pearl River Technology Digital City Loft Apartment
302	天津美年广场 LOFT 办公样板间	Tianjin Meinian Square Loft Model Office
308	台北晶华样板房 C 户型	Regent Taipei Model House C
314	创意 LOFT 里"V 商店"	LOFT Office

008 – 159

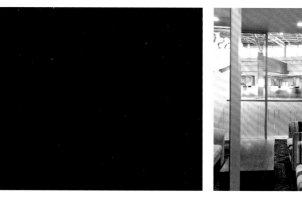

1 Large Office Space
大型办公

Genesis Commercial Building Lobby & Creative Working Space

Genesis 商厦大堂及创意工作区

Design Company: Stefano Tordiglione Design Ltd.
Designer: Stefano Tordiglione
Photographer: David Elliott, Kenneth Tam
Area: 760 m² (Lobby), 1,000 m² (Creative Working Space on the Fourth Floor)

设计公司：Stefano Tordiglione Design Ltd.
设计师：Stefano Tordiglione
摄影师：David Elliott、Kenneth Tam
面积：760 m²（大堂）、1 000 m²（四楼创意工作区）

Creative Revolution in the 1960s and Modernism in the 1970s enlighten the design of Genesis, a business building which was transformed from an industry block in Wong Chuk Hang of Aberdeen, Hong Kong. Being completed in the 1980s, Genesis was designed and decorated by Stefano Tordiglione, and finally we see a brand-new dynamic space including the lobby on the ground floor and a creative working space on the fourth floor.

The space is full of color and vitality. At the entrance of the lobby, features of both Europe and Hong Kong can be seen. On the left wall of the lobby, famous remarks of Bruce Lee are presented by LED lights, and on this wall you can also see Hong Kong street scene in the 1960s. Beside the wall stands an Italian artist's Rolls-Royce artwork, which is an optical device and also an embodiment of this full-of-creation space. Three-primary colors set the main tone of the space. A long wavy sofa and simple lines on the walls add luster the design.

Genesis set people of creative industry as its target tenants, such as artists and photographer. The building is designed to interact with people. Deep in the lobby, a LED screen changes colors when someone approaches. The atrium is two-floor high. Mirrors on the ceiling and walls enable people to see the lobby and themselves from different angles, and chandeliers reflected in these mirrors will bring you to a world of kaleidoscope.

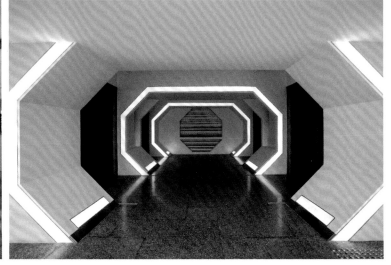

In the center of the lobby stands a five-meter-high palm sculpture with the index finger pointing to the sky, reminding people of Michelangelo's the Genesis, the ceiling frescos in Sistine Chapel, which will help people to associate the sculpture to the creative meaning of the building. Walking up, you will see an octagonal elevator door decorated by neon lamps. Stepping into the elevator hall which is full of sense of sci-fi, you will come to the Creative Working Space that is also a leisure zone.

The fourth floor is in a vivacious style, which stresses the balance between work and life. In this broad public space, furniture is of Italian features. A conference table can be transformed into a ping-pong table, beside which table tennis and a soccer table are available. In a corner of this floor there is the reading zone where stands a wavy bookshelf of 1970s style, and pink lamps will remind people of red droplights on traditional Hong Kong's streets. Two red telephone booths demonstrate the link between Hong Kong and the UK. Walls in the tea zone are decorated by plaid in blue and white. On the carpet, there are newspaper patterns of Hong Kong in the 1970s. Lines and shapes are designed in harmony here, and this idea is also applied in the design of outdoor garden, where scattering furniture forms some cubes, and around the furniture are grassplot and stones. When the lights are on in the night, there is appealing scenery here.

Beside the public space there is a daycare nursery. In the fitness room, placement of the fitness equipments makes it available to enjoy the scenery while taking exercises. A green carpet on the outdoor terrace is arranged for yoga classes.

The design of Genesis is full of creation, and this vivacious space will become a good place to balance work and life.

20世纪60年代的创意革命、70年代的现代主义，启发了Genesis的设计，Genesis是一座位于香港仔黄竹坑由工业大厦改建而成的商业大楼。Genesis于80年代初落成，由Stefano Tordiglione重新进行内部装饰设计。设计后大堂地面及四楼休闲创意工作空间焕然一新，富有动感。

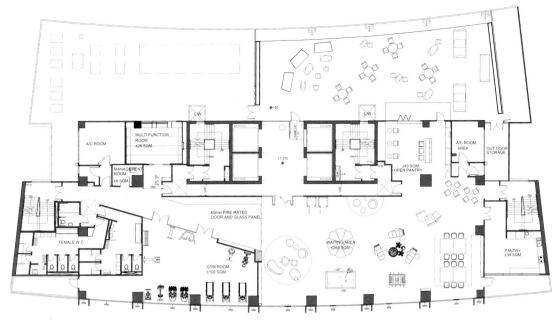

Fourth Floor Plan / 四层平面图

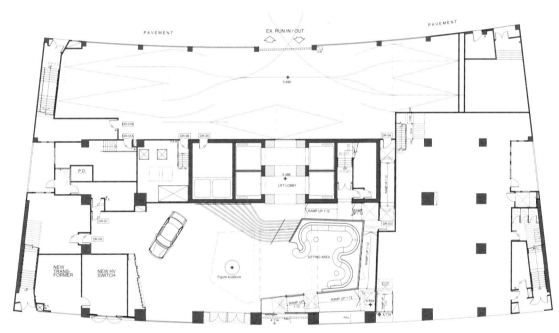

First Floor Plan / 一层平面图

　　整个空间设计色彩缤纷，活力十足。整体风格兼具中西动感元素，大堂入口融合欧洲及香港特色。大堂左边的墙壁以 LED 灯砌出武打巨星李小龙的经典名言，同时亦参考了香港 60 年代满街霓虹灯的独特街景。而在这面墙的旁边，则是一件由意大利艺术家创作的劳斯莱斯作品，它是一件充满灯光幻影的光学装置，同时也是这个充满创意的环境的化身。空间运用了 60 年代常用的三原色，满载设计色卡的基本色调，配衬代表着时代文化特色的长型波浪沙发，整个设计亦因墙上的简单条纹而增色不少。

　　Genesis 以创意工业如画廊、艺术家、摄影师等为目标租户。从大厦的入口开始，设计师的心思便可见一斑，各种元素的糅合，实现了大厦与人的互动。大堂深处安装的交互式 LED 屏幕，每当人靠近时便会自动变色。中庭楼底挑高约 2 层高度，并在墙身及天花上镶嵌镜子作为装饰。镜子以不同的角度摆放，让人们在不同的角度都可以看见自己。大堂中从不同角度悬垂下来的吊灯的倒映，让人仿佛置身于梦幻的万花筒之中。

　　大堂中央竖立着 5 米高的手掌雕塑，其食指直指天空，令人联想起米开朗基罗的名作以及西斯廷礼拜堂天花壁画中描绘的创世纪场景，也正好与这座大厦深具创意的内涵相符。再往上走，便会看见一道以霓虹灯作装饰的八角门，通往充满科幻感的电梯大堂，让人仿佛置身于《2001 太空漫游》之中。这个恰到好处的设计，将大家带到了休闲创意工作空间。

　　4 楼的设计风格是活泼灵动的，强调工作与生活的平衡。公共空间显得朗阔而开放，舒适而不拥挤，家具桌椅的设计灵感源自意大利设计，而会议桌更可转换为乒乓球桌，旁边还设有桌球和桌上足球可供享用。阅读区位于 4 楼一隅，并摆放了可供躺卧及休息的舒适椅子，波浪形的书架则是 20 世纪 70 年代的风格，旁边低垂的粉红色灯，亦使人想起香港传统街市的鲜红吊灯。而其他设计元素亦反映出大厦所在位置的特色。茶水间受 70 年代经典花纹的启发，以蓝白相间的格子作为墙身的装饰元素。地毯上印有香港 70 年代的报纸，细看之下你会发现报纸上报道的都是香港当时的大事件，如海洋公园开幕、香港仔隧道通车等。整个设计透过线条及形状融为一体，如从墙身延伸出吊灯、花盘、电视机及人物的剪影，为大家提供一个休息的空间，同时也可在此工作。而这个设计意念亦活用于户外花园，散落的木制家具组成一个个立方体，其风格亦是源于 70 年代的意大利设计潮流；而这些桌椅四周都是郁郁葱葱的草地和白色的石头，当晚上灯光亮起，景致更是别有一番情调。

　　而在公共空间附近，则是日间托儿室，以 60 年代设计为基调，配以色彩柔

和的彩虹，为办公室的家长们带来便利。健身房也同样经过细心设计，健身器材摆放的方式让人可以一边欣赏黄竹坑道的景致一边运动。户外露台铺设的绿色地毯，则是为室外瑜伽班而特别设计的。浴室不仅备有储物柜、淋浴室等设备，更用充满活力的动感元素加以点缀。

一个专为香港而设计的独立活泼工作空间，Genesis 以其充满创意的设计，启发创新及创造力，带出工作与生活平衡的新思维。

Grant Thornton Office
致同国际办公室

Design Company: Giant Leap

设计公司：Giant Leap

Grant Thornton, the fifth largest auditing, tax, outsourcing and advisory firm in the country, called upon Giant Leap's interior expertise to successfully merge both of their Johannesburg offices under one roof while providing a new and refreshed collaborative work space to further optimize creativity and efficiencies for the group.

The massive 10,800 square meters three storey building positioned at Wanderers Office Park will now be the admired address to more than 470 employees. The company move was also a major milestone for the firm, being the first time in history that the group has become proud owners of the premises.

Commencing in November, importance of the project was placed on uniting staff from two separate locations while providing modern, open plan space and breakaway areas for employees, and ultimately using the Grant Thornton House to showcase the firm's unique corporate identity.

Grandeur, state-of-the-art and stylish—the project outcome was a triumph, but it was not all smooth sailing for the Giant Leap team. Challenges endured included Change Management and staff resistance from individual to open plan collaborative spaces. The project team therefore delivered solutions to merge the two different company cultures with the brand's heritage, and promoted a new way of workspace thinking from its original, traditional approach. Part of this process involved eradicating the traditional office plan by incorporating senior management and partners into a "neighborhood" atmosphere. Giant Leap also incorporated Activity Based Workspace areas whereby staff were assigned an open plan desk with immediate access to quiet rooms, collaboration areas and coffee breakaway zones.

To further promote the firm's culture and different way of thinking, unique design features included an art gallery and the CEO room concept. The art display area on the ground floor exhibits more than 100 pieces by South African artists including; Deborah Bell, Derrick Nxumalo, Helen Sebidi, Judith Mason and William Kentridge. Another design feature, the CEO room, was a Grant Thornton concept which was incorporated to provide space where executives can work on strategic, financial and leadership issues which are free from distractions and operational demands.

致同会计事务所是南非第五大审计、税务、外包和咨询公司。其新办公楼委托 Giant Leap 设计公司进行设计，设计公司以其专业的知识成功地打造了致同会计事务所位于约翰内斯堡的办公室，为员工提供了一个全新的互相协作的工作空间，进一步提升了集团的创造力和效率。

这座三层的建筑有 10 800 平方米，位于漫游者办公园区，拥有 470 多名员工。新办公楼的设计也是该公司一个重要的里程碑，再次翻开了史诗般的新篇章，对公司来说具有极为重大的意义。

设计之初，项目重点在于团结来自两个不同地域的员工，为员工提供现代化、开放式的办公场所，展现公司独特的企业形象。

办公空间富丽堂皇、时尚先进，是办公设计的典范。然而，整个设计过程并非一帆风顺，设计面临的最大挑战则是将管理者的单独办公室纳入员工开放式的办公区中。项目设计团队将两种不同的公司文化和办公特点融合在一起，并从传统的办公方法中衍生出一种全新的办公模式。以"邻居"概念消减高层办公区的紧张感。整体空间设计动静分区，即员工办公区与其他公共性区域及咖啡区分离。

为了进一步发扬公司的文化及思维模式，艺术画廊和首席执行官办公室概念渗透了独特的设计特点。位于一楼的艺术展示区域展览了100多件由南非艺术家创作的作品，这些艺术家包括：黛博拉贝尔、德里克·舒马龙、海伦·瑟比蒂、朱迪斯·马森南·威廉姆肯·翠驰。另一个特别之处就是具有致同概念的 CEO 办公室，为各位高管提供了一个研究战略、财务和领导的良好环境。

Volkswagen Office
大众汽车办公室

Design Company: Giant Leap

设计公司：Giant Leap

Volkswagen Sandton paints the town green: As companies become environmentally conscious - sustainable development has emerged as a growing trend to create high-performance, energy-efficient structures that improve employee comfort and well-being while minimising environmental impact. Now is the time to "Greenovate". Why, because people spend more than 90% of their time inside the artificial environments of buildings and they are starting to expect a lot more from their employers. Volkswagen's Sandton offices were built with innovative "eco" features integrated into the building. Most of the materials used are recyclable with motion sensor lighting throughout. All the furniture was made by a local company called Angel Shack which boosts zero harmful formaldehyde emissions, commonly used in the manufacturing of furniture. Energy efficient air-conditioning was installed and the building is also fitted with highly efficient thick tinted glass for natural insulation and maximum natural light. One of the key features in the building is the indoor hydroponic garden for staff and visitors to enjoy.

大众汽车办公室以绿色环保为设计理念，随着公司环保意识的增强，可持续发展已成为一种趋势，因此办公空间设计以提升员工的舒适度和幸福感为宗旨，同时将对环境影响最小化、能源利用率最大化。在环保创新的时代，人们约90%的时间都身处人工环境的建筑中，并且对雇主的期望越来越高。大众汽车办公室在建筑中融入了具有创新性的"生态"特色，空间中大部分材料都是可回收的，且装满了运动传感照明器。所有的家具都是由当地一家名为沙克天使的公司制作的，保证了有害甲醛的零排放。除了节能空调，还安装了较厚的有色玻璃，这样既可以自然绝缘，又可以实现最大限度的自然采光。值得一提的是，大楼中对员工和访客开放的室内水培花园，是工作之余休闲待客的最好去处。

Beautiful Simplicity Office
朴本原美办公室

Design Company: Yuanshuo Interior Decoration Engineering Co. Ltd.
Designer: Kang Minghua
Area: 160 m²
Materials: Ironwork, Art Stone, Slate, Tawny Glass, Imported Tiles, Wooden Floor

设计公司：远硕室内装修工程有限公司
设计师：康铭华
面积：160 m²
主要材料：铁件、文化石、板岩、茶镜、进口瓷砖、木地板

Our clients, who are two architect designers, ask us to create a "very special" project, and that is all!
On our planning map of the office, there are an achieve room, a storeroom, a large conference room and a staff room, which constructs an orderly and peaceful working atmosphere. To express the pioneering spirit and the feature of "growing out from nothing" in design industry, materials of rough, crude and primary traits, such as tooled-finish stones and rusty slates, are used in the construction of Xuanguan. On the other side of the office, attics can be seen. The office is embellished by bright colors and light industrial style, which echoes the need of earnest and liveliness in the space.
This is an office where crudeness and primitiveness live in harmony with civilization and modernization, which is just the characteristic of the project.

本案业主为建筑师事务所的两位主事者，同时也是两位建筑设计师。因为职业的关系，经常往返于国内外，并且欣赏过许多优秀的作品。因此，在设计上给了我们很大的空间，甚至只给了一个设计大方向，就是一定要"很特别"，其余的就让我们尽情发挥了。

带着两位建筑师的信任，我们开始了本次独特的设计之旅。考虑到空间的性质，即作为公司的办公共享空间，我们在空间功能上规划了事务室、档案室、储藏室，让工作能够在干净整齐的环境中进行。大型会议室，适合各种形式的讨论会。员工休息室，舒缓大家工作的疲累。为突显建筑设计业从无到有的"拓荒精神"，我们使用了能使人感受到粗犷、原始、有开疆辟土氛围的凿面石材、锈板，作为迎接来宾的空间（玄关）。另一边，放眼望去，却是城市顶楼的图像，而位于从无到有之间，这个办公区就像是梦想的孵育区。因此，我们使用轻度的工业风并点缀鲜艳的颜色，为这个需要专注又须保持弹性的过渡区，作了一个很恰当的注解。

这是一个由粗犷、原始，走进都会文明氛围的办公室。两种不同风格的和谐共融，是本案的特色。

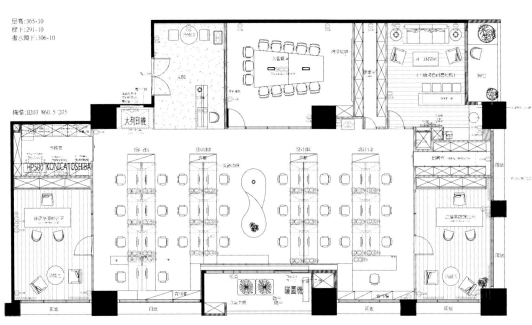

Site Plan / 平面图

HSB Office
HSB 办公室

Design Company: pS Arkitektur
Designer: Peter Sahlin
Participant: Martina Eliasson, Therese Svalling, Emilie Westergaard Folkersen
Phototgrapher: Jason Strong
Area: 9,000 m²

设计公司：pS 建筑公司
设计师：Peter Sahlin
参与设计：Martina Eliasson、Therese Svalling、Emilie Westergaard Folkersen
摄影师：Jason Strong
面积：9 000 m²

HSB is Sweden's largest housing cooperation and owned by its members. It's Stockholm office has just undergone a complete renovation in order to allow for openness and accessibility. pS has been the interior architect and space planner. Much effort has gone into creating an ergonomic and modern office in terms of acoustics and lighting. This in combination with new technology has made the change from cellular offices to open workspace a pleasant experience.

Social interaction and energy has been the keyword and the theme is "Welcome home"! The reception, the so called "Living shop" and the inner courtyard all merge together on the ground floor, allowing for staff and guests to mix and mingle informally. The interior design is comfortable and colorful, contrasting efficiently against the original 40'ies intarsia wall and pater noster lifts.

Some 420 people work in the building. The top floor has an amazing view over the city roof tops and presents a dozen or so meeting rooms for external meetings.

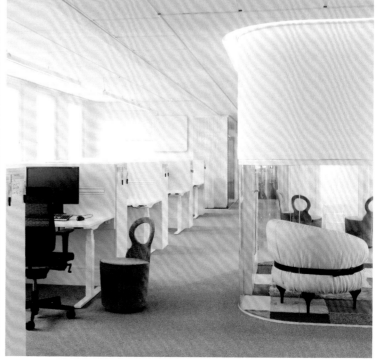

HSB是瑞典最大的房地产公司，为了实现空间的开放性和可访问性，HSB公司对其位于斯德哥尔摩的办公室重新进行了装修与改造。作为主导本案室内建筑和空间规划的pS建筑公司，在营造符合人体工学的现代办公室上，尤其是在音响和照明方面，做出了极大的努力。新科技的融入使空间从传统办公室转化为开放式的工作区，让办公愉悦化。

在空间中，互动与节能是设计一以贯之的关键词，而"欢迎回家"则是本案空间的主题。接待区，也就是所谓的"生活商店"和合并在一起的一楼内院，在这里员工可以进行非正式的接待。与原本无墙、多电梯的接待室相比，该设计舒适而丰富多彩。

办公楼共有420人左右在此办公。在顶楼不仅可以俯瞰全城，更设计了十几个户外会议室。

Digital Agency HQ
Digital Agency 全球总部

Design Company: Steyer Design
Designer: Henrique Steyer
Photographer: Marcelo Donadussi

设计公司：Steyer 设计事务所
设计师：Henrique Steyer
摄影师：Marcelo Donadussi

The clients were looking for something really different and bold to convey the avant-garde DNA of the company, that caters to customers in varied market segments. We created classical walls, with plaster paneling to make a stark contrast with the modern furniture and designer lamps. The Blindfolded Pietá painting, signed by the fictional artist Mark Gary Adams (and created by myself), was on the cover of Brazilian's Vogue Living anniversary edition in 2013 and that was what drew the client's attention to our work.
Signed pieces such as Philippe Starck's Gun Lamp, Frank Gehry's Wiggle Chair and the Dear Ingo lamp are highlights of the decor.
The doors were lacquered in high gloss pink. In the reception, we placed an antique Baccarat lamp that shines alongside the front desk in futuristic style, which was also created by me. High back chairs work as thrones for the visitors.
In the meeting room, the Nemo Chair, by Italian designer Fabio Novambre, creates a hierarchy during business meetings. Its clear that whoever sits on it exerts power over the others, due to the commanding presence of the chair.

本案业主意欲在空间中运用一些既特殊、大胆又能传达公司理念的前卫元素，以迎合不同市场的不同需求。带有石膏镶板的古典墙体与现代家具、灯具设计形成了鲜明的对比。由设计师创作，虚构艺术家马克加里·亚当斯签名的"圣母怜子图"，成功地吸引了客户的注意。

带有签名的菲利普·斯塔克的枪灯、弗兰克·盖里的 Wiggle 椅和 Dear Ingo 灯是装饰的亮点。

接待室里，高光泽的粉色门、大师级的古董百家乐灯、未来主义风格的桌台、奇幻的高背椅等融于一室，设计惊艳而经典。

在会议室，由意大利设计师法比奥创作的尼莫椅子，是权力的象征，只要坐上它马上便会有庄严的感觉，也增强了商务会议室的层级感与严肃性。

NUAC Office
NUAC 办公空间

Design Company: pS Arkitektur
Designer: Peter Sahlin
Architect: Beata Denton
Participant: Mette Larsson Wedborn
Area: 1,400 m²

设计公司：PS 建筑公司
设计师：Peter Sahlin
建筑师：Beata Denton
参与设计：Mette Larsson Wedborn
面积：1 400 m²

Nuac is owned by Swedish Luftfartsverket and Danish Naviair and aims to become the largest providers of air navigation services in Europe in 2012. The office is located high up in the Waterfront Building and offers spectacular views over central Stockholm. The sky is omnipresent and the theme "air plane shape" has decided the choice of furniture and lamps and the specially designed reception desk and wallpaper.
The cantine is the only exception from the theme and instead represents the earth, presenting warmer colors and the feeling of cozy bistro.

Nuac 隶属于瑞典的 Luftfartsverket 和丹麦的 Naviair，旨在成为欧洲最大的空中导航服务提供商。办公室坐落于斯德哥尔摩的海滨建筑的高层，可以俯瞰斯德哥尔摩中央景观，视野极佳。空间中"飞机形状"随处可见，家具和灯具的选择以及特别设计的接待处和壁纸均与主题相适应。

唯一不同的是餐厅，餐厅以地球为主题，着力呈现暖色调，给人以小酒馆的舒适之感。

Shenzhen Jinhui Business Center
锦会商务办公中心

Design Company: Shenzhen Taihe Nanfang Design
Designer: Wang Wuping
Area: 4,300 m²
Main Materials: French Marble, Fire-Proof Plate, Mirror-Finished Stainless Steel, Wallpaper, White Artificial Stones

设计公司：深圳太合南方建筑室内设计事务所
设计师：王五平
面积：4 300 m²
主要材料：云朵拉灰石、防火板、镜钢、墙纸、白色人造石

Smarter Luxurious Space of Jin Club Business Center is a renowned provider of office space in light-luxurious style. It is the preference of some elite industry like innovative finance and innovative technology. Jin Club provides to the entrepreneurs in residence professional business services including financing, evaluation, legal affairs and manpower.
In this project, there is a reception area, a lounge, conference rooms, a VIP room, a drink area, a terrace for leisure, and offices in large and small. The project is in modern simplicity with some creative designs, and the whole space is in an imposing, fashionable and creative taste.

锦会是中国知名的轻奢智慧办公空间服务商，精心为追求卓越的精英人群打造低调、雅致的轻奢级商务空间氛围，是创新金融、创新科技等行业的首选。通过360°智慧服务，锦会为入驻企业提供融资、评估、法务、人力等专业商业服务，用态度和品质开创金融科技智慧商务时代！

本项目在功能划分上有接待区、休闲区、大小会议室、贵宾室、水吧区、休闲露台以及大小不等的多个办公室等。在风格营造上，以现代简约的设计手法，辅以一些灵动的创意元素，让整个办公空间彰显大气、时尚、创意与品位。

Site Plan / 平面图

Design Together
一起设计

Design Company: Crox International Co., Ltd.
Designer: Lin Congran, Hou Zhengguang
Participant: Li Bentao, Yao Sheng, Wang Yantong
Photographer: Hu Wenjie
Area: 1,300 m²
Materials: Cement, Wood, Black Iron, Black Glass

设计公司：阔合国际有限公司
设计师：林琮然、侯正光
参与设计：李本涛、姚生、王琰炯
摄影师：胡文杰
面积：1300 m²
主要材料：水泥、木材、黑铁、黑玻璃

This project is transformed from an industrial factory building of a tall atrium. The chief designer Lin Congran lets stairs of twelve-meter long in the space, which is a simple space arrangement of rituality. The huge stairs and the hall provide a multi-functional reception area for parties, communication leaning and so on. A twelve-meter long passageway will lead you from the wide hall to a bright and spacious working space.

Above the stairs stands a large-scale conference room. A dining room, a library, toilets and the director's office are arranged in the south of the building which is of plentiful sunshine. In the west are a recreational room and a multi-functional room, and other management rooms are in the north of the building. Different zones are divided by glass, so that the daylight is available for the working area. On the second floor, an ambulatory close the atrium connects the first floor and the second floor, where at the end of the corridor a roundabout slide gives an interesting answer of going downstairs.

The building materials construct a simple and natural environment. Different from ordinary office building in the city, this post-industrial working environment represents the sprout of creativity in Shanghai. In this simple cement environment, wooden decoration on the wall and the wooden floor create a warm atmosphere. The cross section of multi-layered plates is the decorative finish of the stairs and corridors. The match of brown glass and black iron work make a modern atmosphere, and a wall decorated by green plants and a fish bowl bring in a natural breath. The harmonious match of different materials in the design achieves the enterprise culture which is "the perfect integration of living and working together".

Lattice system is applied in the design of light to avoid the column shadow in the light. Trunking is cleverly used as the base of LED box, which keeps a clean electric wiring and makes a well-organized modern style with the cement ceiling above. High-light light lines in neutral color installed below

the upper stairs not only illuminate the space but also bring warmth in. In the reading area, the chandelier and projection lamps assembled on the cement columns make three-dimensional effect. There are many delicate design details. The fillet design of columns is not only safe but also pleasing to the eye. The concept of "designing together" and "harmony in diversity" is practiced in the design of toilet. The wash basin in industrial design, different kinds of taps and urinals in various shapes demonstrate the search for multi-element visual enjoyment.

This Designing Together Office is a good model of new creative office, which is an ideal place of growth and communication. All superfluous decoration find no place here. This is a project germinated from the need of users. The crooked finger at the entrance of Designing Together Office delivers honesty, unity and love.

　　设计之初，设计师在面对偌大的挑高工业老厂房时，经过反复地推敲后决定采用一种简单而深具仪式感的空间布局，让长达12米的大阶梯，成为设计概念的主题。这样的空间介质，可以巧妙地实现与新增楼层的过渡，让活动与空间的垂直连通成为可能。巨大的阶梯配合门厅的机能，打造出一个聚会、交流、学习等多功能的接待区域。填入空间的大阶梯视为一种承载了无数活动与记忆的媒介，并在这水平阶梯上嵌入一垂直向度的玻璃量体，直接剖开延伸出12米长的通道。由外而内进入，先由开阔的门厅再进入相对狭小的通道，最终抵达开敞明亮的工作空间，让人在先扬、后抑、再扬的过程中，体会带有韵律感的移动过程。

　　空间机能分布上，在大阶梯的上方设计了大型会议室，让来访的客人感受到那行走间的戏剧性。因此木头大阶梯的存在构成了此地既流动又恒定的日常事件，成为主导空间的精神气质。内部空间的格局，考虑阳光、空气等物理条件，把餐厅、图书室、洗手间与集团总监室设置在阳光最好的南面。西面设置娱乐空间与多功能室，其余的主管室配置于北面，手法上采用开敞与玻璃的分隔方式，让日光直接进入挑空的中央工作区域，在内部建立一个自然的工作环境，利用这样的空间

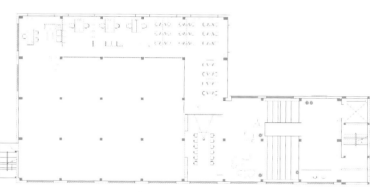

Site Plan / 平面图

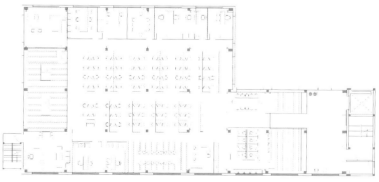

Site Plan / 平面图

规划，清楚界定了一个完整的功能序列，满足了不同部门的需求。另外增建的二楼空间靠近主要挑空区，留设回廊实现上下层间的互动。二楼廊道底端终点的旋转滑梯，是一个极具创意的设计，巧妙地解决了下楼的问题，除了呼应厂房挑高的特色外，还完整地组织出了一种严密的使用罗辑。

在建筑材料的选用上，设计力图呈现出一种朴素的自然构建美学。后工业的办公环境迥异于都市丛林般的写字楼，是一股涌动在上海的勃勃朝气。因此在空间质感的表现上，设计师刻意抹去了过去装修留下的油漆表面，复原了水泥朴素的面貌，另外在空间的地面、墙壁与天花上新添了温润的木作，让具有凝聚力的人文气氛于空间中肆意蔓延。而楼梯和过道采用多层板断层作为饰面，基础的材料叠加在一起最终成为一个更优越的整体，集合断层显示出无数层叠的外观也是借由材料特性暗喻"一起"的美好表达。茶色玻璃与黑铁件的使用，有效产生现代设计感，而通道端景的植生绿墙与装满鱼的水缸更让自然生气涌入；不同肌理的材料搭配出空间的粗犷与细腻，人为与天然的默契，呈现出深具生活质感的整体空间，从而达成设计师最初的设计目标：生活与工作完美融合。

　　灯光设计方面考虑了原始柱网排列现况,采用了格状系统,避开柱子会造成的阴影面,巧妙将走线槽设计成LED灯盒的基座,既避免了凌乱的布线,又与上部裸露的水泥天花构成一种有序的现代风格。在加设的楼板下方贴装中性色温的高亮度灯带,在层高有限的空间内解决了照明问题,间接光的使用也让人感受舒适。重点图书阅览区配设吊灯与水泥柱四边加设立体效果的投射灯,产生空间的层次感,配合挑空的空间产生舞台般的效果。诸多细部设计时也是极尽巧思,如柱子倒圆角处理,兼顾美观与安全。在洗手间的设计上,设计师坚持"一起设计"却"和而不同"的想法,让富有工业设计造型的洗手台,配置上多种类的水龙头及形状各异的小便斗,在使用基础上,积极寻找各种功能和区域间的多元视觉享受。

　　"一起设计"办公室通过了空间整合与改造,塑造新创意办公的使用典范,共同创造出新的设计人生,这是共同成长的空间,也是切磋碰撞的理想所在。正如"一起设计"入口处那个巨大勾手指的标志,时刻传递着真诚,团结和爱。

Shenzhen Archime Co-working Center
深圳阿基米联合办公中心

Design Company: Shenzhen Taihe Nanfang Design
Designer: Wang Wuping
Area: 2,700 m²
Main Materials: Red Brick, Rust Brick, Art Floor, Mushroom Stone, Wood Venner

设计公司：深圳太合南方建筑室内设计事务所
设计师：王五平
面积：2 700 m²
主要材料：红砖、铁锈砖、艺术地板、磨菇石、木饰面

Shenzhen Archime Internet Commune is a cooperation that concentrates on providing the best Internet+ platform for outstanding entrepreneurs. The supporting services provided by them can almost benefit all entrepreneurs, which become integral parts of cities' innovation. All users are cooperative partners with them, instead of lessees, and they will establish a Dynamic entrepreneurial biome together.

In this case, they reconstruct a factory's half storey space, so it only has some unique characteristics of factory, but also some ideas of co-working are implanted in the design. From the plane function planning, we can easily find that by transforming functions antechamber, reception room, leisure room and VIP room into sharing, we can not only strengthen interactions between owners and users, but also users and users. In this way, they can provide more opportunities for entrepreneurs, and share recourses.

In visual forms, the whole room is based on red brick matched with white brick, with some green plant into it, which makes it simple and creative. In the meanwhile, waterscape and bamboo complement each other, being harmonious with pretty mushroom, so the whole atmosphere is combined of emptiness and reality and full of interest.

What stresses most in this design is the leisure place, including coffee bar, Semi-private arc negotiation area, long wooden table for interactive, sofa leisure area, conference chair under hanging bookshelf and so on. These all show the characteristic of co-working, open, interactive, interesting, and relaxing. In this way, they can provide entrepreneurs an environment to contact with each other and achieve win-win finally.

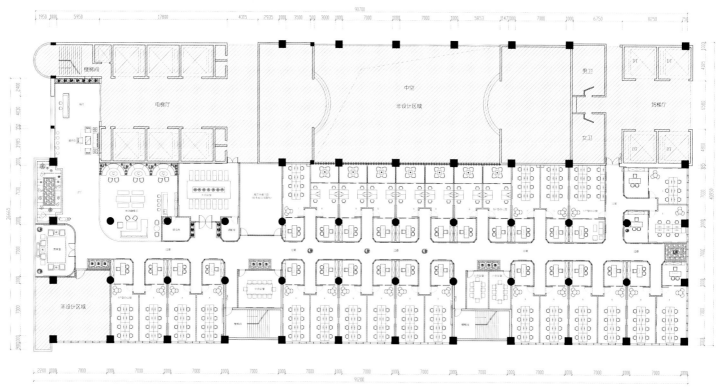

Site Plan / 平面图

深圳阿基米互联网公社，是一家专注于为优秀的创业者提供最极致的互联网+平台，他们提供的配套，不仅服务于园区企业，更为所有创业者所共享，成为城市创新的有机组成部分。他们所有的用户，不是租户，而是合作者，将共同构建创业社区，打造一个动态的创业生态圈。

本案是由一栋厂房的半层空间改造而成，设计后的空间仍保留了一些厂房特有的属性，并将联合办公的服务理念植入其中。从平面功能规划中不难看出，将共用的前厅、接待室、贵宾室、休闲区、会议室等空间，从传统的办公功能系统中剥离出来，在实现资源共享的同时，加强了业主与用户、用户与用户之间的互动，从而为创业者提供了更多的契机，实现资源共享。

在设计视觉形态上，红砖元素贯穿全场，并在其中饰以白色的造型框和绿植，空间简洁、创意，韵味尽现。营造洽谈气氛的水景和创意视觉的干竹互融互生，相得益彰，又和旁边硬朗的蘑菇石构成和谐关系，轻重有度，虚实有别，惟妙惟肖。

休闲处是本案设计的重点，设有咖啡吧、半私密的弧形洽谈区、长木互动桌、沙发休闲区，以及小型吊装书架下的双人洽谈椅等，这些无不彰显联合办公的开放、互动、趣味与休闲融为一体的办公特质，最终实现创业者之间的相互交流、资源整合、合作共赢。

D3 Interactive Environment
D3 交互式的环境

Design Company: Estudio Guto Requena + ilo Design
Participant: Lucas Ciciliato, Paulo de Camargo, Diego Spinola, Olavo Spinola
Photographer: Fran Parente
Area: 50 m²

设计公司：Estudio Guto Requena + ilo Design
参与设计：Lucas Ciciliato、Paulo de Camargo、Diego Spinola、Olavo Spinola
摄影师：Fran Parente
面积：50 m²

D3 is a youthful and vigorous company. This company likes to pursue new concepts and develop new areas. This company is specializing in website and procedure development. So, an office space with creative ideas and interactional function was needed urgently. It is a great challenge for the designer to create a space which could inspire the design concepts and realized the interaction between the staff and the space. This area is 55 square meters. The designer attempt to realize the interactional function and present a unique style. It is a big challenge. However, the designer keeps the original architectural functions. The ceiling was reserved its former decoration. It was simply painting with nude oil paint. The caisson ceiling with steel structure enhanced the industrial style to the whole space. The wooden furniture was more delicate. The space colors were expertly planned. The dark and light blue, green and purple were presenting a rational and calm sense. It is conformed with the company needs. The design of the furniture reflected the flexibility and interaction of this space. The wheeled chair meets the different needs from the staff.

D3 是一家热衷求知、年轻而有活力的公司，尤其喜欢尝试新鲜的事物、涉猎新的领域。其工作内容以开发网站与应用程序为主，迫切地需要

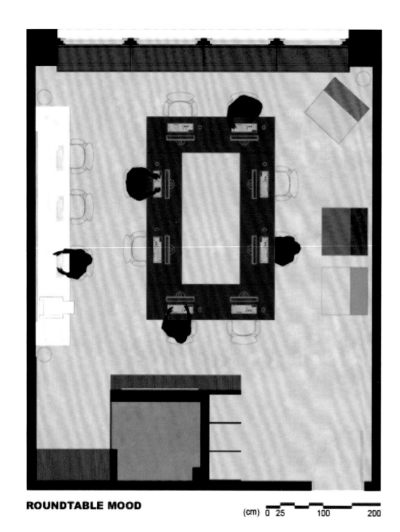

ROUNDTABLE MOOD (cm) 0 25 100 200

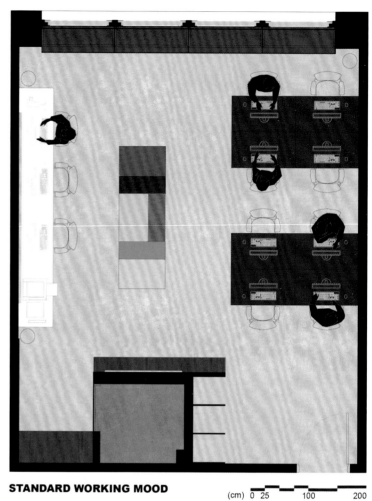

STANDARD WORKING MOOD (cm) 0 25 100 200

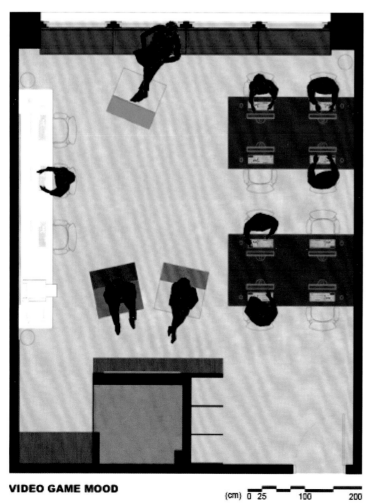

VIDEO GAME MOOD (cm) 0 25 100 200

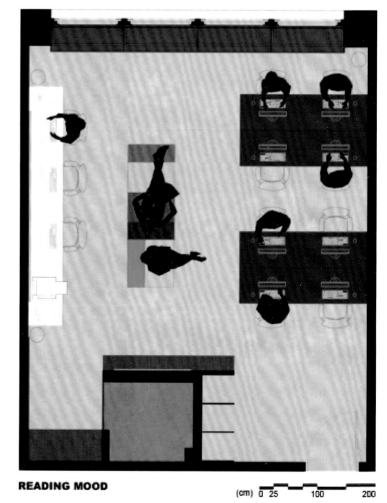

READING MOOD (cm) 0 25 100 200

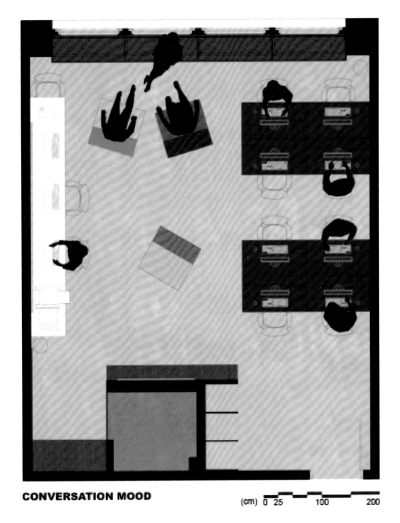
CONVERSATION MOOD　(cm) 0 25 100 200

PRESENTATION MOOD　(cm) 0 25 100 200

WATCH TV MOOD　(cm) 0 25 100 200

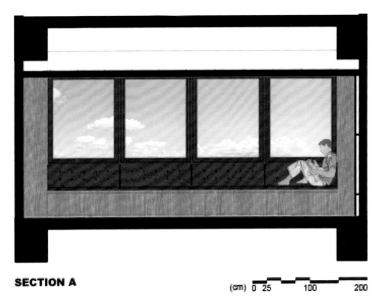
SECTION A　(cm) 0 25 100 200

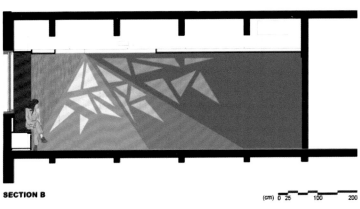
SECTION B　(cm) 0 25 100 200

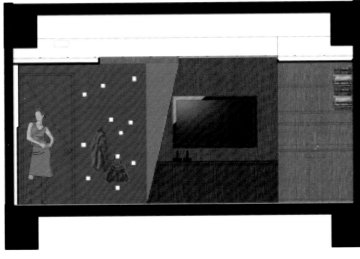

SECTION C

(cm) 0 25   100   200

有一个集创意性、趣味性与交互性于一体的办公空间。因此,对于设计师来说本案最大的挑战就是如何去创造一个可以激发员工创作灵感,实现员工与员工之间、员工与空间之间的互动关系的空间。

　　本案面积仅有55平方米,在这样一个小空间中既要实现空间主体的多元互动,又要呈现独具一格的空间面貌,对于普通人来说是一个不小的难题,但设计师匠心独具,极尽设计之手笔。首先设计师对建筑原始功能进行了保护,如天花并未进行过度的装饰,简单地刷上一层裸漆后暴露于人们的视野之中,天花上藻井式钢结构更增强了空间工业风格的意味。同样在地面之上,设计师去除了老旧的地毯,将地板上的树脂层显露出来。与粗糙工艺的天花与地板相比,部分精心设计的木质家具,则显得更为精致与高端。其次,空间色调的选用上也颇具匠心,不同深浅的蓝色、绿色和紫色理性而沉稳,也正好符合了公司工作内容的需要。最后,在家具的设计上高度地体现了空间的灵活性与互动性,可移动的滑轮椅可满足不同情况的使用需求。

Quadria Capital Singapore Office
Quadria Capital 新加坡办公室

Design Company: Elliot James Pte Ltd.
Designer: Elliot Barratt
Photographer: Ken Tan
Area: 335 m²

设计公司：Elliot James Pte Ltd.
设计师：Elliot Barratt
摄影师：Ken Tan
面积：335 m²

Designing a luxurious Hedge Fund office space in Singapore, Elliot James were briefed to create a stylish, comfortable office that whilst having a very high end finish, would be inviting and informal and take on the look and feel one would associate with a New York apartment combined with an English Gentleman's club.

The project site was situated on the upper floors of three combined shop houses, which provided an unconventional space due to the original architecture of the building. This presented the opportunity to be extremely innovative with the interior design layout, and allowed the design consultancy to create a variety of different break-out areas where staff could remove themselves from the intensity and immediate stresses, even if only momentarily. These spaces include a secluded library, hidden breakout area, generous lounge, pantry, boardroom, and private offices.

Wanting to be sensitive to the architectural features of the building, Elliot James created private rooms and offices that preserved the open plan space by way of glass walls and partitions with thin black framework to separate the various rooms. Original features such as the thick concrete columns and exposed beams were left exposed. The classic shop house windows were highlighted by a contrasting black finish and the original concrete flooring was stripped back, polished and varnished to a high marble-like shine.

The interior design firm created a simple panelled wall for the main office and bespoke bookcase in rich black for the library area.
Through a mixture of lighting, Elliot James were able to create a soft, inviting ambience with areas of low lighting to help relax and escape from the more intense areas of the office.
The furniture was designed with deep, rich colours in mind in various quilted leathers and velvets. Using fabrics from Timorous Beasties, Elliot James designed a custom made bench with mirrored gold legs. The coffee table was also bespoke designed by the ID consultancy and finished in the same mirrored gold with smoked black glass for the table top. The specific tone of gold was to be an accent colour used throughout to conjure that feeling of luxury, opulence and sophistication.

本案是一个位于新加坡的投资公司办公室，由设计师艾略特·詹姆斯主导设计。设计定位为时尚、舒适、高端，让人们从外观便联想到纽约的公寓或英国绅士俱乐部。

办公室位于三个商铺的上层，从原本的建筑结构来看本案是一个非传统的空间。室内空间布局极具创新意味，各种不同的场景，让员工可以从高强度的工作中解放出来。办公室设计有隐蔽的图书馆、隐藏的扩展区域、休息室、厨房、董事会和私人办公室。

为了能让人们很好地感受到建筑的特色，设计师艾略特·詹姆斯在设计会客室和会议室时，运用带有黑色薄框架的玻璃墙来实现空间的开放式设计。原来厚混凝土柱和暴露横梁的保留，使空间略显原始与粗犷，但也增添了些许别样的韵味。古典商店式黑色窗户的采用点亮了整个空间，赋予空间一种古典的气息。原始的混凝土地板也被剥去，取而代之的是经过抛光、涂漆的闪亮大理石。

办公区域设计中，设计师以一面简单的玻璃格子墙，既分隔了空间区域又使空间敞朗开阔。图书馆区域定制的两面深黑色书柜，整齐的相对而立。

通过混合照明设计，设计师在低照度的区域营造了一种柔软而又有魅力氛围，在这里员工可放松心情，释放工作压力。

在家具的选用上，设计师选用了深色的皮革与天鹅绒家具。来自Timorous Beasties的织物，设计师亲自设计的带有镜子的金腿长椅，由ID咨询公司定制设计的咖啡桌，桌面为黑色玻璃的金腿镜面茶几等，以及定制的调和金色作为主色调，给人以豪华、富裕和成熟的感觉。

095

SAB Miller India
印度 SAB Miller 办公室

Design Company: Zyeta
Designer: Shilpa Revankar, Kishore M/Anita Yadagiri, Anil Bhardwaj
Photographer: Mani Iyer
Area: 4,459 m²

设计公司：Zyeta
设计师：Shilpa Revankar、Kishore M/Anita Yadagiri、Anil Bhardwaj
摄影师：Mani Iyer
面积：4 459 m²

SAB Miller is one of the world's leading African brewery companies with their presence in over 80 countries. SAB Miller has grown from its original South African base into a global company with operations in both developed markets and in emerging economies such as Eastern Europe, China and India.

Design inspired by the rhythm and color pallet, which signifies the vibrant and enthusiasm of space. The warm, solid color, patterns and classic design to create a cohesive environment to achieve optimum work productivity and occupant satisfaction. The goal is to create harmony and unity through the use of colors and patterns. The linear pattern and light accents are intended to create an energetic feel. A variety in the spaces keeps the employees excited about the work place, and keeps them focused, Colorful carpets and liner battens breaks the monotony and creates a compelling spot.

SAB Miller 是世界领先的非洲酿酒公司之一，其生产销售遍布全球80多个国家。SAB Miller 酿酒公司已从南非公司发展成了一个具有影响力的跨国公司，其市场占有包括东欧的发达国家以及中国和印度等新兴经济体。

设计灵感来自于节奏和调色盘，彰显着空间的热情与活力。温暖、纯色、模式和经典设计营造出一个具有凝聚力的办公环境，极大地提升了工作效率和使用者满意度。

本案期望通过颜色和模式的改变来营造空间的和谐与统一之感。线性模式和照明旨在创建一种充满活力的办公空间。多样化的设计提升着员工的热情和专注力，彩色地毯打破了空间的单调，为员工和访客呈现出一个变化的趣味空间。

Corporate Office for Maxim Integrated
Maxim Integrated 办公室

Design Company: Zyeta
Designer: Rakshith Kumar, Anil Bhardwaj, Krishna
Photographer: Mani Iyer
Area: 3,252 m²

设计公司：Zyeta
设计师：Rakshith Kumar、Anil Bhardwaj、Krishna
摄影师：Mani Iyer
面积：3 252 m²

Maxim Integrated is an American, publicly traded company that designs, manufactures, and sells analog and mixed-signal integrated circuits. The Brief was to create a unique space with more privacy still collaborative types for 250 seats along with five Individual Labs with Different features.
Design inspired from the Process of Manufacturing the silicon wafers. As a basic process, Silicon is purified in multiple steps will have a bright vibrant colors during the process and finally reaches semiconductor manufacturing quality which is called Electronic Grade Silicon.
Vibrant color Cubes represent the Ion implementation process. The exposed areas of the carpet are bombarded with various accent color carpets.
Natural plants and recycled materials are introduced all throughout which induces a nature's tinge at the working environment and also relishes the professional image.

Maxim Integrated 是美国的一个上市公司，专门设计、生产和销售混合信号集成电路。本案则是为其设计的一个独特而具有较强隐私性的办公室，该办公室需要有协作办公类型的办公席位250个，以及具有不同特性的实验室5个。

本案的设计灵感来源于硅片的生产过程。作为一个基本过程，硅提纯有多个步骤，在此过程中硅将会呈现一种美好而充满活力的色彩，最后制成的半导体，即电子级硅。

空间中充满活力的方形色块即代表了离子变化过程的颜色变化。地面上铺设的缤纷地毯，使空间更为绚丽与清新。

空间中大量天然植物的运用，为员工营造了一个具有自然气息、生态环保的工作环境，而可回收材料的采用也恰到好处地彰显了公司的职业形象。

Japanese E-commerce Company
日本某电子公司

Design Company: Zyeta
Designer: Shilpa Revankar, Anita Yadagiri, Nithin PL
Photographer: Mani Iyer
Area: 910 m²

设计公司：Zyeta
设计师：Shilpa Revankar、Anita Yadagiri、Nithin PL
摄影师：Mani Iyer
面积：910 m²

We ensured to consider everyone's need; we discovered that the people wanted more small discussion rooms, Telephone Booths, Brainstorming and Breakout areas. We designed the flexible workplaces to allow employees to sit anywhere they wish to.
we have designed all the work spaces towards large windows to allow lot of natural light to create comfortable and open office. All closed spaces like Meeting/Discussion Rooms are towards solid areas. The acute angle of this floor is fully glazed, highly visible and is the best view of Bangalore. We created the variety of seats: high seats, low seating, movable seats, and lounge seats creating different moods.
Ceiling is kept open at the workstation area to make space visually higher. All the ducts are kept open but painted grey matching to ceiling color. Rest of the area is kept white and plain.

设计最大限度地考虑到了每个员工的使用需求，尽可能多地设计小房间、电话台、洽谈区。办公场所设计灵活性强，员工可随意地坐在任何地方办公。

办公室所有的工作空间都设计有大幅玻璃窗，将大量的自然光线揽入室内，使办公室更加舒适、开放。凭窗远眺，眼前是班加罗尔壮观的城市景致，视野极其开阔。所有会议室、讨论室等封闭空间的室内布局相对固定，地板采用的是过釉瓷砖，并以各种类型的座位，如高座位、低座位、可移动的座椅、休息室座椅等，营造不同的空间氛围。

员工工作区是一个开放式的办公区域，灰色的天花与管道裸露于空间表面，工厂风格的装饰简约而自然。在色彩的运用上，灰色的天花彰显空间的低调与宁静，偶尔跳脱出的红色则为空间增添了些许活力。白色的墙面、办公桌与黑色办公椅的搭配，总能让人凝神静气，专注于手中的工作。

easyCredit HQ
easyCredit 总部

Design Company: Evolution Design
Photography: TeamBank AG / EasyCredit, Christian Beutler
Area: 15,000 m²

设计公司：Evolution Design
摄影：TeamBank AG / EasyCredit, Christian Beutler
面积：15 000 m²

The biggest challenge was the change from a traditional workspace with dedicated desks to a completely new activity-based style of working. Thus, the building was divided into, Homezones' and Meet & Create Zones. Each Homebase' consists of a unique mixture of different workplace typologies to cater for the different needs of the employees depending on their daily activities.

The internal functional organization of the building utilises the idea of a city: a main staff restaurant and a Barista Bar with a vibrant marketplace' feeling, the Homezone', consisting of several Homebases' creating individual local neighborhoods for the employees to work in, and the inner ring of the building as the Meet & Create Zone'. All these public areas are part of the city concept with parks' and street scenes' to encourage a higher frequency of informal and accidental meetings.

Another very important part was the identity and branding concept. Throughout the building all spaces were designed in order to bring the company's culture and vision alive. The result is a building in which easyCredit and its employees can find an inspiring home which reflects their extraordinary company spirit and enables them to identify strongly with their new workplace and HQ, and in which they can share their values of fairness, teamwork, transparency and communication everyday.

本案最大的挑战是要将传统的工作区与专门的部门转变成一个以工作性质为基础的活动空间。因此，空间被分为家庭地带、会议和创意地带。每个家庭地带包含着不同工作类型的独特组合，根据员工的日常活动来满足其不同的需求。

建筑的内部功能布局源自于城市的构想：员工餐厅和咖啡酒吧给人以活力之感。由几个家庭基地组成的家庭地带，专为员工打造。建筑的内圈是"集合创造区"，所有的公共区域都是城市概念的组成部分。公园式、街景式的环境有利于员工进行高频率的非正式会议。

整个办公空间设计的目标是让公司更有活力，体现公司文化与经营理念。在这里，员工似乎回到了自己温馨的家中，放松、自在的工作环境，也充分反映了公司的企业精神，让员工具有强烈的认同感和归属感，也体现了公司公平、透明、沟通的价值观。

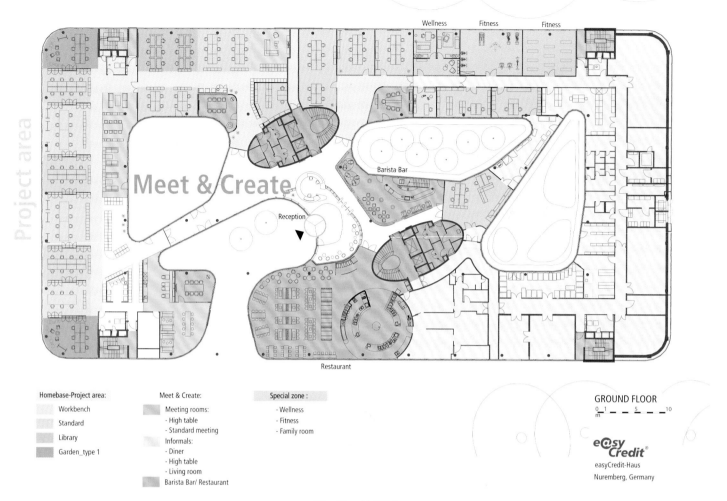

Site Plan / 平面图

133

BBDO Indonesia Office
BBDO 印度尼西亚办公室

Design Company: Delution Architect
Designer: Muhammad Egha, Hezby Ryandi, Sunjaya Askaria, Fahmy Desrizal
Participant: Naufal Ryandi, Sigit Widigdo
Photographer: Fernando Gomulya
Area: 900 m²

设计公司：Delution Architect
设计师：Muhammad Egha、Hezby Ryandi、Sunjaya Askaria、Fahmy Desrizal
参与设计：Naufal Ryandi、Sigit Widigdo
摄影师：Fernando Gomulya
面积：900 m²

BBDO is an agency engaged in the field of large-scale Global advertising based in New York. When this project tendering brief began, the client wanted a very different and dynamic office shades which can provoke creative ideas from the employees. Thus we made a very unique, eye-catching and philosophically creative design that is applied to almost all the elements in the interior design of this Indonesia BBDO Office. Starting from the work table, floor, furniture, lights, doors, wall elements, to the lighting concept, each area has a unique and different lighting concept with warm white tone. The lobby area has a big wave lamp made of corrugated iron put together in a wavy form, giving the flow impression for the people entering the office.

Other than these lights, visitors will also be greeted by Indonesian-themed receptionist table with batik stamp elements that are processed into the table leather in order to assert that BBDO Indonesia also has a local character even though it is based in New York.

After passing the receptionist area, visitors will encounter a communal space composed of puff chairs shaped like a ladder that can be set in many ways. Behind this area is the main boardroom hidden between three-colored decorative wood elements which completely hide the boardroom and as main wall who prop the main BBDO slogan signnage.

The working area uses a box concept that is applied by parquet wood and clear glass, making an impression that there is a room inside a room thus the leaders may still have privacy while maintaining the performance of his subordinates.

Each room in the BBDO Office Area has different lighting concept. The Boardroom

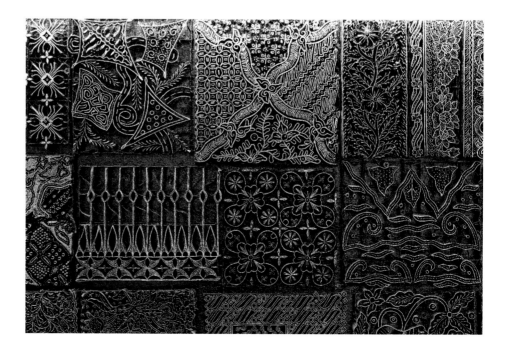

uses continuous lighting concept which is integrating wall and ceiling lighting elements. Small meeting rooms use square concept with different sizes arranged randomly.

The working area uses Maze-shaped lamps arranged from basic TL lamps. The Creative Director Area uses raw light bulbs arranged in different heights to philosophically represent the ideas of the creative director.

Also, some spots have small meeting tables with floor-sitting concept imitating the Japanese culture such that each employee can discuss without having to use the meeting room.

All corners of the wall can be written to arouse the generation of creative ideas. This concept is delivered by using black glass and blackboard. The whole office uses Contemporary Industrial concept where all the materials are still unfinished, e.g. the polished concrete floor and ceiling showing off all its utility elements.

BBDO 总部设在纽约,是一家大型的全球性广告公司,本案是其位于印度尼西亚的办公室。因客户想要一个与众不同的动态办公室以激发员工的创意,所以设计师做了一个非常独特而富有哲学意味的创意设计,将各种元素应用其中。从办公桌、地板、家具、门、墙面到照明设计,每个区域都采用独特的暖白色照明灯。大堂天花别具一格,暖白色的灯光镶于波纹形状的铁件上,并

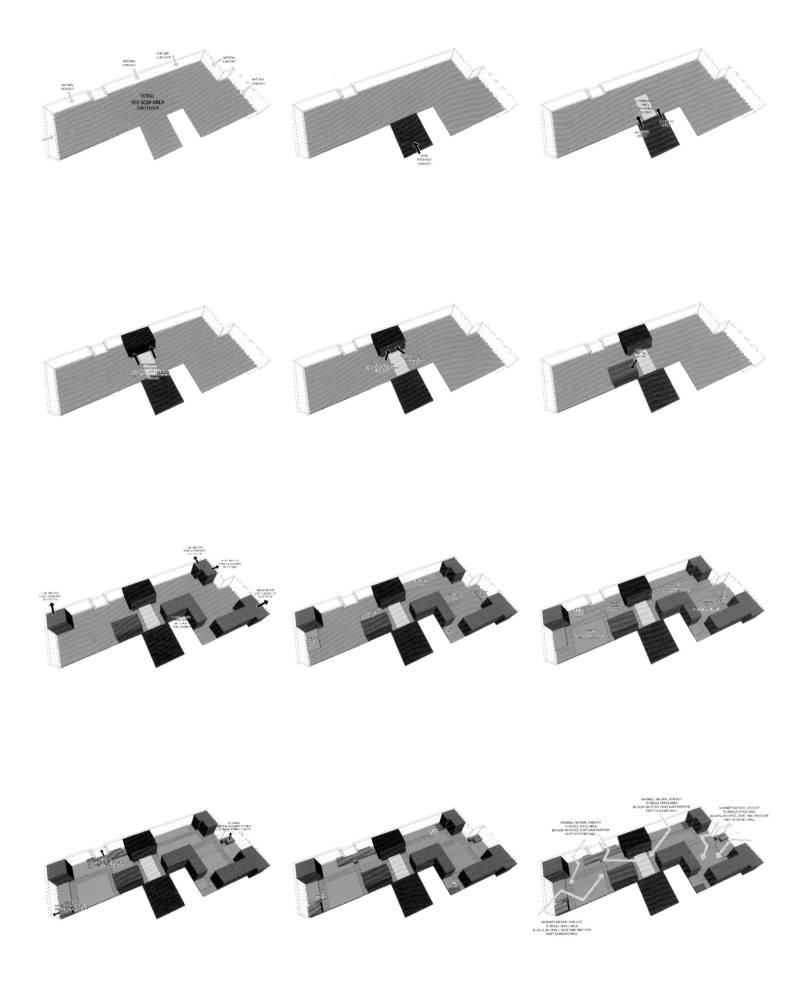

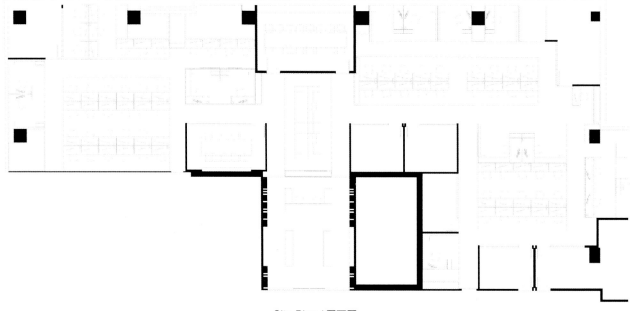

Site Plan / 平面图

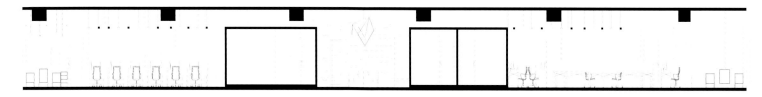

Sectional Drawing / 剖面图

以波浪形式有序铺陈，让进入大堂的人们感觉空间似乎是流动着的。

除了这些富有特色的照明设计外，访客也能欣赏到印度尼西亚特色的蜡染印花工艺，它们被应用于前台正面的皮革之上，以彰显 BBDO 印度尼西亚公司的地域特色。

经过前台区域，呈现在访客眼前的是以各种形式展现的梯形椅子所组成的公共区域。公共区域隐藏在作为公司标志墙的前台白色背景隔断墙之后。

工作区的领导办公室以玻璃盒子的概念进行诠释，给人一种屋中屋的感觉。在这里，领导既可以监督其下属的工作又保护了空间的隐私。

BDDO 办公室中的各个空间的照明设计各有特色。会议室采用了连续性的照明概念，灯光从墙壁一直延续到天花之上，极富魅力。小会议室里则随机地安排着不同大小的方形灯具。

工作区迷宫式的天花上，镶嵌着最基本的 TL 灯具。创意总监区域使用高低不同的原始灯泡，哲学般展示着创意总监的创意。

除此之外，还设计了带有类似日本会议桌形式的非正式会议空间。不用会议室，员工在这里也可自如地进行工作讨论。而且每个通往小会议桌的阶梯下面的空间皆可当成储物空间。

墙体的每一个角落都镶嵌有黑玻或黑板，员工可以用白色奇异笔在玻璃或黑板上挥洒其创意思维，激发创作灵感。从整体上看，办公室使用的是当代工业风格，所有的材料都裸露在表面，如抛光混凝土地板和天花板。

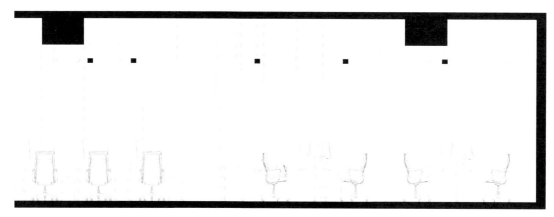

Sectional Drawing / 剖面图

Site Plan / 平面图

T2 Headquarters
T2 总部办公室

Design Company: Landini Associates
Photographer: Trevor Mein
Area:1,300 m²

设计公司：Landini Associates
摄影师：Trevor Mein
面积：1 300 m²

Showcasing the creation of a new life for an old building, Landini Associates brief was to restore the former Collingwood industrial warehouse into the new headquarters base for T2. The aim was to respect and celebrate the original building, while creating a dramatic statement to represent the T2 company's ethos. Throughout the building, timber beams, columns and brick walls were sandblasted to bring back their original finish whilst a few scuffs and marks are kept as a remembrance of its history. This was contrasted with T2's trademark dark, streamlined pallet.

Perceived constraints were turned into an advantage, exploiting various levels within the warehouse while designing a unified space. This was due to the site being located on a sloped block through to the rear street and resulted in the creation of different arrival experiences for both the public and staff. With the street frontage lower than the general entry, the public is welcomed into the space through an oversized steel framed pivot door into a double height space. Led by concrete steps, guests are greeted by an oversized T2 logo lit by incandescent light bulbs. To the left of the entrance is a Tea Bar/reception, an informal place to stop, drink and chat to guests, personifying T2's identity. This space also acted as the trial concept for T2's latest tea-retailing concept T2B, which similarly features a large cast concrete bar and Tea Library.

The staff entry is from the upper level rear car park, from where they walk down onto a catwalk in the centre of the office warehouse. This acts as the sites backbone, yet separates the space into two, housing different functions of the business. Both sides of the catwalk feature different faces. One, a clean facade featuring sleek black steel panels that hide the catwalk, while the other with a timber panels concealing storage systems.

While the rawness and industrial experience is continued upstairs in the tea making, tasting workshop and CEO's office, a completely different color palette

is employed. Here a white, clean space encourages light and nature to enter the room through the opened windows.

The development marks a new era for T2, providing an innovative and functional office space that reflects a deep understanding of culture, urbanity, and of course, celebrates the centuries-old art of tea-making.

　　Landini 建筑设计公司将前科林伍德工业仓库重新装饰设计为 T2 新总部基地，为人们展示了一个具有新生命力的老建筑，表达了人们对老建筑的尊重与纪念，同时也戏剧性地表明了 T2 公司的宗旨。整个建筑的木梁、柱体和砖墙都进行了喷砂处理，但仍保留了原建筑的刮擦处和斑点，这些斑点也成了建筑的历史纪念品。T2 公司的商标设计采用明暗对比的手法，也是现代化设计的体现。

　　设计师对仓库的不同层次进行统筹规划，设计出了一个统一的空间，曾经的限制性因素，如今却转化成了优势。由于仓库位于一个通往后街的倾斜板块上，临街的那面低于一般的入口，访客进来时，要通过一个超大的双层高度的钢框架枢轴门。走过混凝土台阶，就能看到一个超大闪亮的白炽灯 T2 标志。左边的入口是一个茶吧/接待处，这是一个临时性休闲场所，在这里可以喝下午茶、接待来访客人，极为人性化，也增强了 T2 员工的认同感。大型现浇混凝土吧台和茶库相似，正呼应了 T2 的最新的茶叶零售 T2B 理念。

　　工作人员进入到上一层后部的停车场，从那里走过一个步行小道到达办公室仓库中心。这作为公司的后端，却又将空间分为两半，刚好隔开了不同的业务功能，而步行小道两侧各有特色，一边是刚好隐藏小道的干净立体的光滑黑色钢面板，而另一边是隐藏存储系统的木板门。

　　在意犹未尽地参观完工厂的制茶车间后，走进工作区和首席执行官的办公室，却发现这里别有洞天，空间色调与其他区域完全不同，整体以白色为主，彰显空间之素雅，光线顺着开着的窗户飘洒进来，既提供了良好的自然采光与通风，也将窗外优美的自然景致揽入室内。

　　设计为 T2 公司提供了一个功能齐全的创新性办公空间，也暗示着 T2 公司将迎来一个全新的时代。设计是对文化和都市的深入解析，更是对历史悠久的烹茶艺术的展示。

Coop Office
Coop 集团办公室

Design Company: pS Arkitektur
Designer: Peter Sahlin, Pernilla Wass
Participant: Beata Denton, Anna Johansson, Viktor Ahnfelt

设计公司：pS Arkitektur
设计师：Peter Sahlin、Pernilla Wass
参与设计：Beata Denton、Anna Johansson、Viktor Ahnfelt

The task was to create a logic but generic office layout with the capacity to adapt to future changes of activities and functions. We had the role of contract managers responsible for all fixed interior decorating and furnishings as well. By creating an open and transparent office, the corporate identity of Coop was strengthened as well as the staff's sense of belonging. We have been aiming at a modern and ergonomically sound working environment that caters to the individual employee's needs.

该设计项目的任务是创造一个有富有逻辑性且通用的办公空间，以适应未来活动变化和功能变化的需要。同时项目安排了专门负责所有的室内装饰和家具装饰的设计师。通过开放透明的办公空间的营造，提升 Coop 集团的企业认同感，同时强化员工的归属感。设计师倾力打造符合人体工程学的现代办公环境，以满足员工的个性化需求。

160-283

2 Creative Office Space
创意办公

Havas/Arnold Worldwide Boston Headquarters

汉威士集团 / 阿诺国际传播波士顿总部

Design Company: Sasaki
Area: 11,613 m²

设计公司：Sasaki
面积：11 613 m²

Havas/Arnold Worldwide is a leading global ad agency that represents major brands like Fidelity, Jack Daniels, Progressive Insurance, and a host of other household names. Havas's new headquarters in downtown Boston have centralized Arnold Worldwide with three other Havas brands. Together, these brands cohabitate on three and a half floors in the former Filene's building, designed in 1912 by renowned architect Daniel Burnham. Sasaki was engaged to lead the design and build-out of the new office space, completed in 2014.

Authenticity and innovation are core to Havas's creative work. Accordingly, the Sasaki team melded the building's architectural authenticity with future-forward design by employing two concepts. The first, the stitch, links the north and south sides of expansive floor plates, tying together old brick, terra cotta, and steel with glass, concrete, and technology. The reuse of found materials—old railings as marker trays, radiator grilles as light fixtures, reclaimed wood for stair treads—and restrained use of new materials speaks both to the agency's dedication to authenticity and to its mission of sustainability. The second concept, the vortex, creates a vertical circulation and collaboration core from the roof skylight to the lowest level, anchoring common amenities around central staircases.

The design supported Havas's goal to shift from 85% private offices to a completely open floor plan. Workstations line the perimeter, surrounded by natural light and views. Here, leadership and employees sit side by side. To aid the reorientation to an open-office plan, the project team ensured the availability of alternative workspaces. Employees can engage in organic moments of exchange, whether at the cafe, two office bars, sky-lit mezzanine, staircase seating areas, or hidden nook. Every square foot is active, engaged, and useable—even circulation zones are lined with writable surfaces, doubling as alternative meeting spaces.

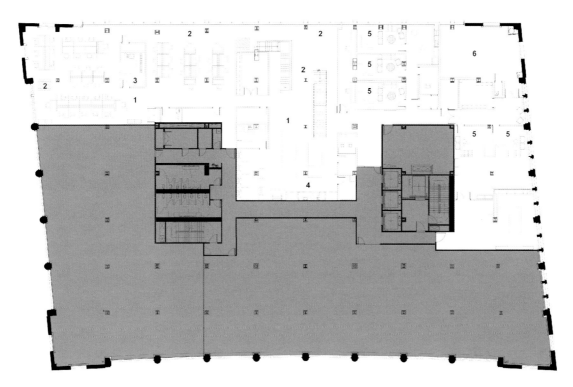

6th Floor
1. Open Office
2. Open Collaboration
3. Phone Room
4. Workroom
5. Edit Room
6. Shoot Room

Sixth Floor Plan / 六层平面图

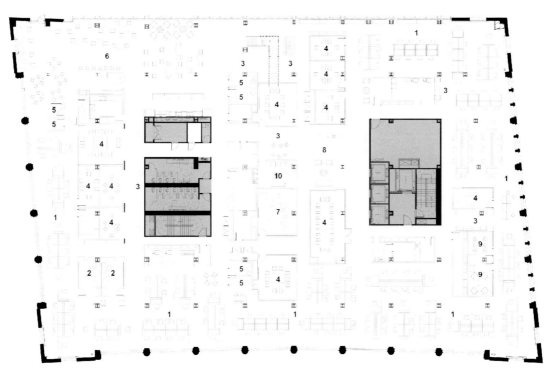

5th Floor
1. Open Office
2. Office
3. Open Collaboration
4. Conference Room
5. Phone Room
6. Creative Commons
7. Training
8. Lobby
9. Team Room
10. Emerging Tech

Fifth Floor Plan / 五层平面图

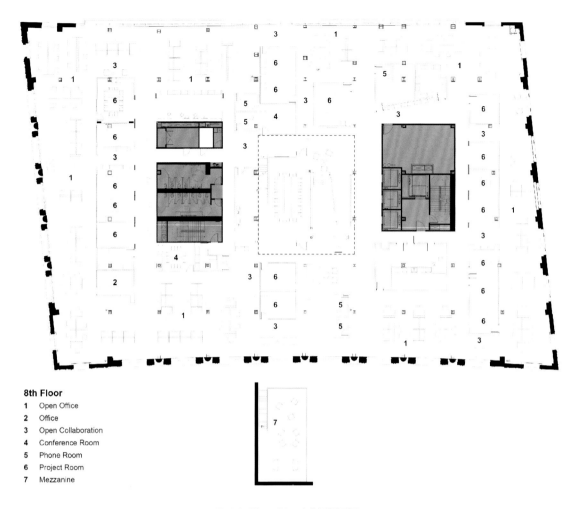

8th Floor
1. Open Office
2. Office
3. Open Collaboration
4. Conference Room
5. Phone Room
6. Project Room
7. Mezzanine

Eighth Floor Plan / 八层平面图

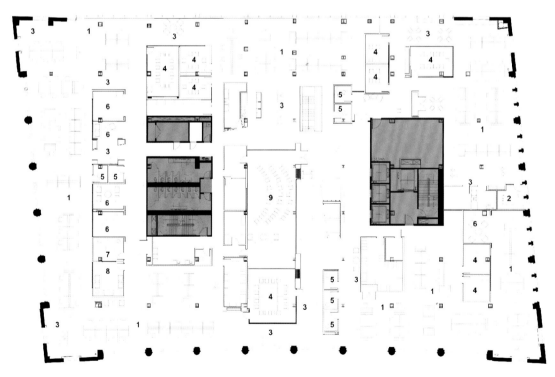

7th Floor
1. Open Office
2. Office
3. Open Collaboration
4. Conference Room
5. Phone Room
6. Project Room
7. Viewing Room
8. Casting
9. Town Hall

Seventh Floor Plan / 七层平面图

The project team worked very closely with Havas/Arnold Worldwide to help create a space that stitches together the history and tradition of Boston—embodied in the Burnham-designed building—with the creativity and innovation of Arnold and its people. The team also worked with the developer and architect of the new onsite residential tower to utilize a split core instead of the originally designed central core, helping the floor plate lend itself to a more fluid design.

汉威士集团/阿诺国际（Havas/Arnold Worldwide）是一家全球领先的广告公司，代理包括美国富达国际、杰克丹尼、美国前进保险公司等家喻户晓的品牌。汉威士集团在波士顿市中心的新总部集中了阿诺国际与旗下其他三个Havas品牌，占据了由著名建筑师丹尼尔·伯纳姆（Daniel Burnham）在1912年设计的前法林百货（Filene's）建筑中的三层半空间。Sasaki主导了这个新办公场所的设计，并于2014年顺利完工。

真实性和创造性是汉威士集团创意工作的核心。因此，Sasaki团队运用了两种设计理念，将对原建筑的尊重和前卫的设计手法完美结合起来。首先，连接南北两侧延伸的地面，尝试将旧砖、赤陶土、钢材和玻璃、混凝土与技术进行结合。对现有材料的重新利用，如把旧栏杆改造成笔架、散热器格栅改造成灯具，再生木材改造成楼梯踏板等；以及减少对新材料的使用，都体现了业主对原真性的追求和推动可持续发展的使命。其次，设计创建了一个垂直循环的协作核心贯通屋顶天窗和最底层，将公共设施设置在中央楼梯的周围。

该设计帮助汉威士集团完成了由85%封闭式办公环境转化为开放式办公空间的计划。办公桌沿外围布置，可尽情沐浴阳光并享受窗外景色。在这里，领导和员工并肩而坐。考虑到办公空间被定位为开放式，项目团队同时还设置了一些替代性工作空间，譬如咖啡厅、两个公司酒吧、有天窗采光的阁楼、阶梯式座位区和隐蔽式工作区等，供员工随时随地进行交流。设计保证每一寸空间都是有活力的、可用的，甚至流线区都采用了可以画写的表面，可兼作会议空间。

设计团队与汉威士集团/阿诺国际密切合作，在这个伯纳姆设计的历史建筑内，将波士顿的历史传统和阿诺国际品牌和其员工的创新精神完美结合。设计团队还协同开发商和该楼住宅项目的建筑师，把原有的中央核心筒分开，从而保持了设计的流畅性。

Victory Hangzhou Branch Office
百利文仪杭州分公司办公室

Design Company: ADD+ Creative Organization
Designer: Guo Weicheng
Area: 1,400 m²

设计公司：ADD+ 创意机构
设计师：郭为成
面积：1 400 m²

We are devoted to provide the clients a joyful office experience, but not only a furniture product.

To create a flexible, ecological, energy-efficient and environmental office space is the conception of dynamic space, and the trend of office environment nowadays. Therefore, this dynamic space reflects the value of Victory just right. The "dynamic" in "Dynamic space and joyful work" connotes "environmental protection, low-carbon and green", and "joyful" demonstrates the pursuit of quality office life. Victory advocates "efficient, joyful, innovative and free" office life, which is just the office life and environment of Victory.

There is a change. And the change in design industry is that the work of professional designers is participated by proprietors. It is important for the modern offices to harmonize the design of furniture and environment in the office.

"The professional of dynamic space builder" is the company's orientation. It's our idea that designing offices with industrial decoration instead of traditional decoration. The designers integrate six elements, namely, the sky, the earth, the wall, the light, the sound and the taste, to make a harmonious environment for people in the offices.

该企业致力于为客户创造一种愉悦办公空间体验，而不仅仅只是为客户提供一件办公家具。

设计为客户创造灵动、生态、节能环保的办公空间环境，是"活态空间"的设计理念的体现，也是当下办公趋势的要求。所以，这个活态空间恰好是百利的价值所在。"愉悦办公"中的"活态"体现了环保、低碳、绿色等重要内涵，"愉悦"更是在办公环境中追求的办公生活化。百利集团倡导高效、愉悦、创新、自由的办公品质生活，在其办公空间中营造这样一个办公环境再合适不过了。

设计引领价值改变，设计也正由专业设计师的工作向更广泛的用户参与演变，

以用户为中心的、用户参与的创新设计日益受到关注。然而在办公家具界，只有办公家具设计和办公环境相通才是现代办公空间的需求。

百利以"活态商务空间整体方案解决专家"作为企业的定位，以工业化装潢替代传统装修的理念来改造未来的办公环境，以融合了天、地、墙、光、声、味六位元素设计理念，使新办公空间下的人与环境和谐共存。

Masisa Office
Masisa 办公室

Design Company: Estudio Guto Requena
Designer: Guto Requena
Participant: SP Estudio, Tatiana Sakurai
Photographer: Fran Parente
Area: 300 m²

设计公司：Estudio Guto Requena
设计师：Guto Requena
参与设计：SP Estudio、Tatiana Sakurai
摄影师：Fran Parente
面积：300 m²

For the Masisa Brazil headquarters in Sao Paulo, Estudio Guto Requena created an interior design identity based on analyses of the company and interviews. We chose a color chart with neutral tones such as grays and beiges, punctuated with green accents. At every step we sought solutions that relied on furniture that exemplifies Brazil, such as in the reception area which includes items from the renowned national designers Fernando and Humberto Campana, Sérgio Rodrigues, Paulo Biacchi and Jader Almeida.

We used Masisa wood panels in a palette of light colors throughout the office, around the ceilings, in the cabinetry, and in the bathrooms and service areas, demonstrating the product's variety of uses. For the ceiling we created a pattern that distributes wood plaques diagonally in a way that lends a strong identity to the space and integrates all the work areas. This same pattern is reflected in the muted tones of the carpet tile flooring, cut to the same size as the ceiling plaques.

The lighting was designed to strengthen the interior identity of the Masisa headquarters, with a dynamic zigzag arrangement of utilitarian fixtures suspended from the ceiling. In the reception area we sought a more indirect and diffuse lighting that creates a more welcoming and cozy ambiance. We also designed a neon lamp for this area that gives a surprising and less corporate aspect to the entrance. The Masisa logo on the wall here greets visitors, while also functioning as a showcase of wood veneer plaques in an artful arrangement that refers visually to pixels on a screen. All the plaques on this screen of pixels are detachable, enabling the customization of this wall over time and allowing architects and designers to examine and familiarize themselves with Masisa products on the adjacent work tables.

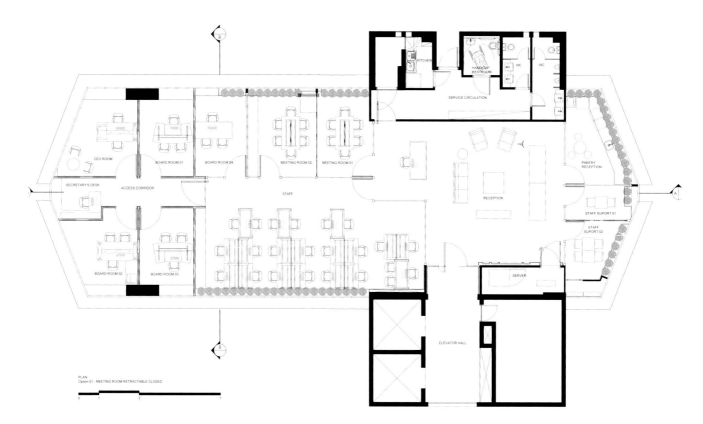

Site Plan / 平面图

Sectional Drawing / 剖面图

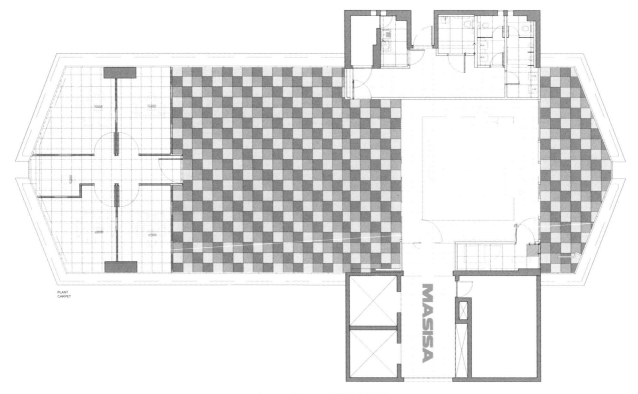

Carpet Layout / 地毯布置图

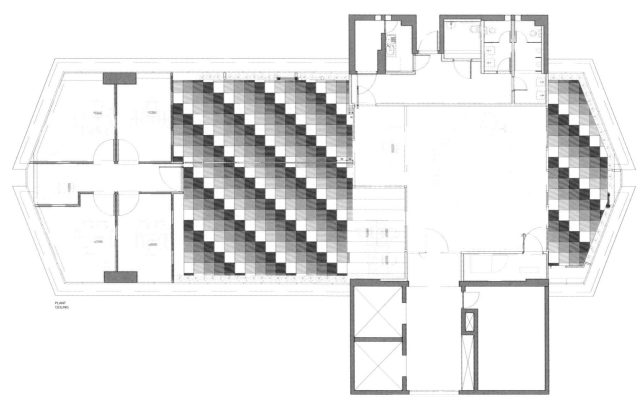

Ceiling Layout / 天花布置图

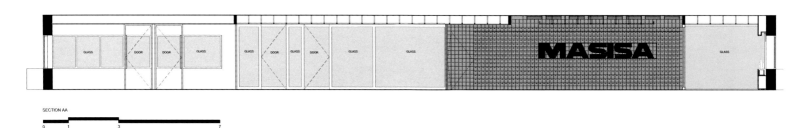

Sectional Drawing / 剖面图

Flexible space was prioritized in the work areas by using mobile dividers, and cabinetry and furniture on castors that permit, for example, the meeting rooms to be rearranged in different configurations as the need arises. Each executive office was customized according to the preference of colors and furnishings chosen by the directors themselves.

A green belt of seedling Snake Plants cuts through the office to bring a natural element of Brazilian character to the design, and to remind us that offices these days do not need to appear bureaucratic, cold or devoid of emotion. All material choices in the interior design of Masisa's Sao Paulo headquarters strive to reduce environmental impact, utilizing domestic and certified products assembled with local labor.

Masisa 公司巴西总部设在圣保罗，Estudio Guto Requena 工作室基于对公司的分析和研究，为其创建了一个全新的公司形象。设计以中性色调为主，如灰色、米色，并以绿色进行点缀。在空间设计中，家具是最能体现室内风格与特色的，而本案中选用的家具则完美地体现了巴西的风情，如接待区的家具就有巴西著名设计师 Fernando、Humberto Campana、Sergio Rodrigues、Paulo Biacchi、Jader Almeida 的作品。

整个办公室，如天花板、橱柜、卫生间和服务区，其产品功能多样化，而办公室门使用的是浅色调的 Masisa 木板门。在天花板上，设计保留了木板上的斑斑点点，给人一种很强的认同感，同时使办公区域形成了一个整体。同样铺在瓷砖地板上的暗哑色调地毯，也裁剪得跟天花板上的斑块一样大小。

Masisa 公司总部的天花板上稳固地悬挂着 Z 字形排布的动态灯具，这些照明设计都强化了内部形象设计。在接待区，安装了朦胧的漫射照明，营造了愉悦温馨的氛围。在这一区域，还设计了一个霓虹灯，在进来时就能让人叹为观止，而少了一种进入企业的感觉。访客步入空间，墙壁上 Masisa 公司的标志即能尽收眼底，木质的饰面上投影的标识巧妙地展示了公司的形象。而随着时间的变化，墙上的投影内容是不断地更新转换的。

工作区的设计，优先考虑灵活性，通过在文件柜和家具上安装脚轮而确保其可移动性。例如，会议室可以根据不同的需求而重新布局。每个行政办公室的主管可以根据自己的喜好定制空间色彩和办公家具。

一条种满虎皮莲的绿化带横穿办公室，浓郁的巴西风情弥漫空间。在这里，装饰设计对环境的影响降到了最低，所用材料也都经过了环保认证。

Hongwen Space
鸿文空间

Design Company: CEX Hongwen Space Design Co., Ltd.
Designer: Zheng Zhanhong, Liu Xiaowen
Photographer: Liu Tengfei

设计公司：CEX 鸿文空间设计有限公司
设计师：郑展鸿、刘小文
摄影师：刘腾飞

To design an office seems, sometimes, to make a box. It is the designers' aim to give people who come here a chance of forgetting time and settling down to think in this simple and natural place.
There are few gorgeous materials, only cement and paint can been seen. The terrace brings most outdoor scenery into the interior, and to see though a frame on the wall in the interior, the faraway mountain is like a landscape painting. Green plants are here and there in the office. Therefore, whether to look out from the interior or to look into the office outside, you are always staying in a quiet and beautiful landscape.

设计办公室有时候就像在打造一个盒子，设计的宗旨是希望不管是工作在里面的设计人员，还是委托设计的业主，只要一进入这个空间，仿如只要穿过一个窄小的盒子，就从尘世穿越到了超然世外。在光影交错且朴实自然的空间里忘掉了时光，真正静下心来思考。

本案没有用过多华丽的材质，整个空间只用水泥、油漆，让淡然宁静的光影隐匿在朴实的水泥里。于露台处，空间最大化地利用了户外景致，且在露台正面用框的形式把室外的远山框进了室内，宛如一幅山水画横放在空间里。室内的空间更设计了多面绿色的墙植，不管是在室内看室外，还是在室外看室外，都似乎处在一幅安静的人文山水画里，品一杯香茗，点一株檀香，任思绪飘飞，让整个人完全处在空灵的世界里。

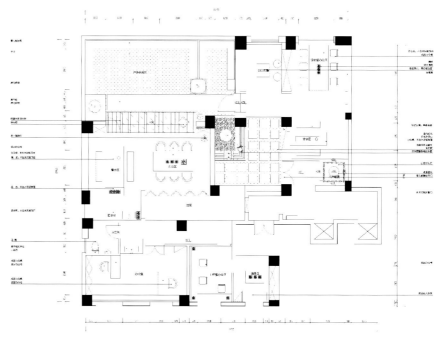

Site Plan / 平面图

Infocomm Investments (BASH) Office
Infocomm 投资办公室

Design Company: SCA Design (a member of the ONG&ONG Group)
Project Director: Brandon Liu
Designer: Johannah Posa, Belnice Chua
Photographer: Bai Jiwen
Area: 2,221 m²

设计公司：SCA Design (ONG&ONG Group 成员之一)
项目总监：Brandon Liu
设计师：Johannah Posa、Belnice Chua
摄影师：Bai Jiwen
面积：2 221 m²

The new office of Infocomm Investments, also known as BASH (Build Amazing Startups Here), is a fun and vibrant space that adapts quickly to the needs of its users. Centrally placed hubs are highly visible, encouraging gatherings and interactions, while specially designed workstations can be adapted to accommodate future growth and expansion. To inspire users and visitors, the overall design is vibrant, lively and fun with an industrial edge, featuring colorful signage and themed areas throughout the space.

The client, Infocomm Investments, builds and invests in Singaporean and global Infocomm technology start-ups to accelerate their growth. In building a vibrant and sustainable ecosystem for start-ups to flourish, they needed a space that would support training programmes, co-working environments, as well as networking events for visitors and occupants. Hence, their brief called for flexibility, mobility and adaptability of the space, as well as an environment that would motivate and inspire budding start-ups to develop their business ventures.

The team addressed the client's requirements by designing workstations that can be arranged in different ways to support various team configurations and workplace activities. The contemporary design demonstrates space optimization through movable space dividers, privacy screens, as well as custom built "mobile booths" that can be adapted into collaboration spaces

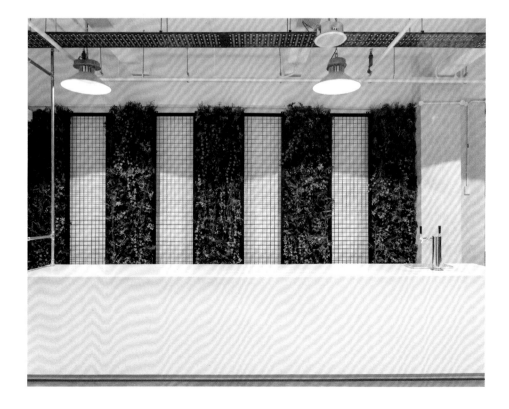

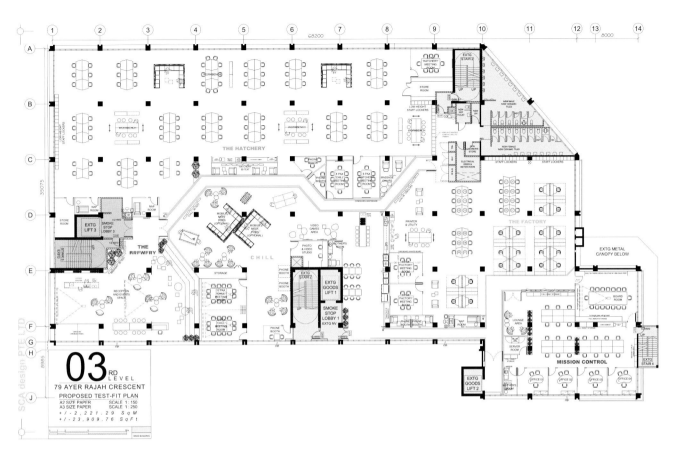

Site Plan / 平面图

or meeting areas. In addition to folding worktops and adjustable shelving cabinets, the designers have devised modules on castor wheels that feature open shelving, writable and corkboard surfaces (mobile booths) for users to segregate space easily as an alternative to the open office space. By having office furniture that is easily adaptable to suit their needs, occupants can use the workspace more effectively in accordance with the fast-paced world of Infocomm technology.

BASH contains five different zones, each with its own brightly color-coded overhead cable-trunking to help visitors to find their way around the space. Each zone is announced with the bold use of large signage that easily distinguishes the various areas from each other whilst maintaining the open concept look. Visitors first encounter The Brewery, which is a reception area that features "stadium seats" set against the wall to maximize the holding space within the zone. Following that is The Hatchery, a co-working environment featuring custom-built movable workstations; Chill, a shared breakout space with recreation elements; and The Factory, another co-working environment that features customizable workstations and "mobile booths". The fifth zone is Mission Control, where Infocomm Investments sits.

Together with maximised light entry to the entire space to allow for natural illumination and views of the outdoors, these elements create fun and comfortable work environment. This is further enhanced by the quirky and playful design of the cafe and chill out areas, which can be used for informal discussions. This open and conducive environment is inspiring and functional, thus providing young start-ups with a vibrant, socially engaging and intelligent workplace that reflects the needs of its users.

Infocomm投资公司的新办公室也被称为"BASH"，它是一个有趣而充满活力的办公室，能快速满足用户的需求。办公空间中央的设计是集中而紧凑的，鼓励着员工团结互爱互助，而专门设计的工作站能够适应未来的发展和扩张。为了给员工和访客深刻的印象，空间设计主题覆盖了整个办公区，产业元素生动有趣，引导标识丰富多彩，室内设计充满活力。

Infocomm投资公司随着业务的发展，需要一个能满足培训需求、团体办公，以及能为访客和工作人员提供各种配套服务的办公空间。简单地讲，也就是要一个灵活多变、可移动及充满激情和斗志的工作环境，去激励员工奋进开拓。

设计公司为满足客户要求，设计出了能够按照不同的方式来开展各种团队和工作活动的空间。设计以可移动的隔断、屏风、定制办公桌，作为洽谈室或会议室，实现空间利用的最大化。除了折叠桌和架子文件柜外，设计师已经对脚轮进行模块功能设计，可以轻松地用开放式架子、可写软木板（移动办公桌）

进行办公空间分隔。办公家具的设计极大地满足了员工的工作需求，在这里员工可以更有效地工作，跟上 Infocomm technology 公司快节奏的工作生活。

BASH 包含五个不同的区域，每个区域的架空电缆的颜色都不一样，鲜亮醒目，便于访客们找到自己寻找的区域。每一个区域都有大型的标志，这就能很容易地区分的各个领域，同时给人一种大胆开放的感觉。"啤酒厂"区是一个接待区，里面设置了"体育馆式的座位"，最大限度地保持区内的空间，这样的设置前所未有。接着就是"孵化场"区，这个共同的工作区是定制的，且可以移动。"寒气逼人"区，它是一个娱乐细胞爆发的休闲区域。"工厂"，是公司的另一个工作群区，是可定制的可移动工作站。第五是"地面控制中心"，是 Infocomm Investment 的控制中心所在。

整个空间的设计尽可能多地引入户外的自然光线，同时也最大化地利用室外景观，拓宽景观视野，营造出充满趣味性和舒适性的工作环境。而奇特俏皮的咖啡厅和休息区，用于非正式的接待是最好不过了。这种开放且有益于身心的工作环境，不仅鼓舞人心而且功能性强，为年轻企业提供了一个充满活力、斗志昂扬的工作场所，满足了员工的需求。

Midea Forrest City Times Model Office
美的·林城时代办公室样板房

Design Company: C&C Design Co., Ltd.
Designer: Peng Zheng
Participant: Chen Jitian, Chen Yongxia
Area: 110 m²
Main Materials: Stoving Varnish Plank, Carpet, Black Mirror Steel, Glass

设计公司：广州共生形态设计集团
设计师：彭征
参与设计：陈计添、陈泳夏
面积：110 m²
主要材料：烤漆板、地毯、黑镜钢、玻璃

Being located at the CBD of future city center of Guiyang, Midea Forrest City Times consists of shopping malls, leisure mall streets and first-class office building.
To meet the demands of different target guests, three office models are designed, which is the big one, the medium one and the small one, which is the type of this project. Being designed in a simple and lively style, the office of a small media company contains creativity and affinity.
Lines intensify spatial tension of the office that is mainly in white. A young and lifeful quality can been seen in the delicate reception, the distinguished ceiling and a carpet of vivacious patterns. The cupboard door is made from glass of backing enamel finish. It is a surprise to see a reading and viewing area between the shear wall and the outside wall.
You will find the passion, happiness and dreams of starting up a business here.

美的·林城时代是美的地产在贵阳注入巨资倾力打造一个重点项目，整体规划由大型商场、休闲商业街、超甲级写字楼组成，位于贵阳未来城市中心CBD的核心地段。

本次设计针对不同的目标客户群分别设计了大中小三套办公室样板房，本案为其中的小户型，以小型文化传播公司为背景，整体设计简洁明快，在有限的空间中体现创造性和亲和力。

横向拉伸的线条贯穿于白色的主色调中，强化了空间的张力。轻巧的前台、独特的天花、跳跃的地毯，这些都体现出空间年轻而充满活力的气质。活动柜门被设计成可涂写的焗漆玻璃，最让人惊喜的是设计师将原建筑剪力墙与外墙之间的狭窄区域设计成一个可以观景的阅读区。

窗明几净的会议室，简洁明快的总监室，纯洁的白和青葱的绿，以及无限的都市天际和天边的一丝云霞，似乎让我们看到了创业的激情、快乐和梦想。

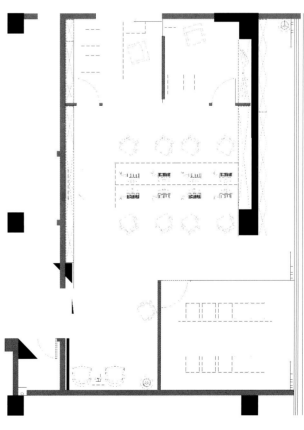

Site Plan / 平面图

Payguru Office
Payguru 办公室

Design Company: MIMARISTUDIO
Designer: Ayca Akkaya Kul, Onder Kul
Photographer: Alp EREN
Area: 210 m²

设计公司：MIMARISTUDIO
设计师：Ayca Akkaya Kul、Onder Kul
摄影师：Alp Eren
面积：210 m²

Payguru Office is a technology office situated in Ari Teknokent 1, which is the first technopark building in iTu Teknokent complex. It is 210 square meters in area and its interior design, project design and site application were all provided by is mimaristudio team.

Payguru is new generation payment platform which gives services like in-app, mobile and subscribed payment also meets all regulative requests. The firm went into an inside company transformation to adapt changing country and business conditions. They decided to start this change from their physical structure which is their office.

Dynamic structure of the sector which Payguru is active and the young staff profile whom will be the new designs main users became the starting point for the design project. The liveliness in the interior architecture style and in the usage of materials in the finishes are references to these aspects of the company.

The usage of daylight is prioritized in the planning process so that all of the workplace could benefit from natural light. After the relations between the different internal units of the firm are studied, the divisions between units are decided to be landscape elements. Especially the direct relationship between the office space and the garden boosted the landscape applications that are designed for the interior.

When visitors enter the place, they arrive at a custom made welcome desk after passing through the entrance hall. By this desk area, where the general feeling of the space is captured, there is a hosting place which can be used not only by the visitors but also by the employees. The area which is diversified with a game console is also associated with the service kitchen nearby.

The main space which is right after the entrance area is an open workspace with a balanced planning. The managers were not excluded from the "open working

space" approach due to the companies' inner organization. Separation is only emphasized in the groupings of the furniture and in the colors used. Also with this approach, senior management working units are solved in a semi open space on an 8 meter working line that also serves as a meeting table. In addition to these, a semi closed meeting room is designed for closed meetings.

Mimaristudio, as their general approach to design, has selected simple building materials that are less in variety and are complementary to the general concept. On the floors, 5 mms of PVC tiling is used upon the raised floor which is rendered to reflect the technological face of the firm. For regulating effect of the spatial geometry of the place to acoustics, panels in varying sizes and colors that are complementary to the design concept are used after the calculations. These acoustic panels are placed in a random order with the lighting fittings forming a complex surface geometry.

Mimaristudio prioritizes the consistency with the general design concept in the design and selection of movable and fixed furniture. While dynamic products with high technology are selected for the office chairs, tables with sharp lines are preferred for the work groups. For file cabinets, products specific for the spatial concept are designed. These products which vary in 3 different colors and different dimensions are more than usual office furniture.

The lighting design of the building is also done by the mimaristudio team as an extension of the spatial design concept. LED technology is chosen for the entire buildings lighting works. Professional collaboration with Crealux firm is made for the lighting consultancy to calculate requirements for different lighting levels for different locations of the project. Decorative lightings are also installed in addition to lighting luminaries which compose the general lighting design. Graphic design applications and visuals are also applied in the interior design likewise other projects done by mimaristudio.

Payguru办公室坐落在Ari Teknokent 1——一个科学技术办公楼中，是iTu Teknokent建筑群的第一个科技园。办公室面积达210平方米，它的室内设计、工程设计等都是由Mimaristudio团队提供。

Payguru是新一代的支付平台，它给为应用程序、移动支付等提供支付服务，订阅支付也在其服务范围内。该公司进入了一个内部转型阶段，为适应不断变化的国情和商业环境，公司决定从内部开始进行转变，即着手改变其办公环境。

Payguru是充满活力的，办公空间的设计则以年轻员工为考虑的重点，采用活泼的室内建筑风格及与之相适应的装饰材料。

在规划的过程中，优先考虑日照的问题。所有的工作场所都有充足的自然光线，空间宽敞明亮。在研究了企业内部各单元间的关系之后，各单位之间的划分主要考虑景观因素，特别是办公室空间和花园之间的直接关系，注重将室外景观引入室内。

当访客进入这栋楼，再通过入口大厅就到了一个有定制接待桌的区域。通过接待区，你就能感受到整个办公楼的独特之处，这里不仅可以用来接待访客，内部员工也可以使用。这个区域还新增了一个控制台，能远程呼叫附近的餐厅服务。入口处右侧区域是一个开放式的工作区。

由于公司内部结构问题，管理者是很乐于接受"开放式工作空间"的设计的，设计师们只会在家具的组合和颜色的使用方面重点突出空间的分隔。采用同样的方法，高层的办公区设在一个有8米工作线的半开放的区域中，这个区域也可以作为一个会议区。除此之外，专为封闭式会议而设有一个半封闭的会议室。

与一般的设计概念刚好相反，设计师选择的建筑材料品种较少，而且较为简单。活动地板砖用的是5毫米厚的聚氯乙烯系地砖，充分体现了企业的高科技风貌。为了调节空间几何结构限制造成的隔音效果，经过计算之后，隔板的大小和颜色的设计都是不同的。这些隔音板随机地与照明灯具搭配安装，形成了一个复杂的几何形状。

Mimaristudio工作室在设计和选用移动和固定家具时会优先考虑常规的设计，如线条明晰的办公桌。而文件柜则是专为适应空间概念而设计的产品。文件柜有3种不同的颜色与大小的办公家具。

建筑的照明设计是空间设计概念的延伸，也是由Mimaristudio设计团队完成的。整个建筑的照明灯具采用的是发光二极管技术。同样，其他的项目也是由Mimaristudio团队完成的。

Municipality Eindhoven Office
埃因霍温某办公室

Design Company: M+R Interior Architecture
Designer: Hans Marechal, Bart Diederen, Nena van Gemert, Marcel Visser
Photography: Studio de Winter
Area: 2,100 m²

设计公司：M+R Interior Architecture
设计师：Hans Marechal、Bart Diederen、Nena van Gemert、Marcel Visse
摄影：Studio de Winter
面积：2 100 m²

Modern organizations are continuously evolving, and a core concept is flexibility: being adaptable and pliable. In a metaphorical sense of the word this can be applied to the character of a person, organization or building, symbolizing the ability to easily adapt, should the situation demand it.

Individuals are able to work anywhere, with the help of the continuous development of communication aids. Fixed office spaces no longer fit in this picture. The "new way of working" demands a different work environment. For example, just think about organizing the workspace based on activities: concentrated work in a quiet room, consultations is a meeting room, brainstorming in an inspiring room, and even working from home can be included in this concept. The flexible configuration of workplaces offers organizations the additional benefit of using spaces more efficiently.

For this concept to work in practice, the interior design needs to offer as much support as possible for all the new functions, without making the configuration too inflexible. Yes, it's all about flexibility. Green offices, sustainable buildings, from cradle to cradle concepts which are of interest.

The attention to these themes is constantly increasing and the frameworks for projects are defined in terms of durability. Basis for sustainable building is designing a good living and working environment with a low energy consumption so we can decrease the use of limited resources.

In addition to the positive effect on nature, the environment and surroundings are especially the installation concept and the choice of material that affects the indoor environment in offices. Research indicates that improving the indoor environment a increases in labour productivity by 5%~10%.

217

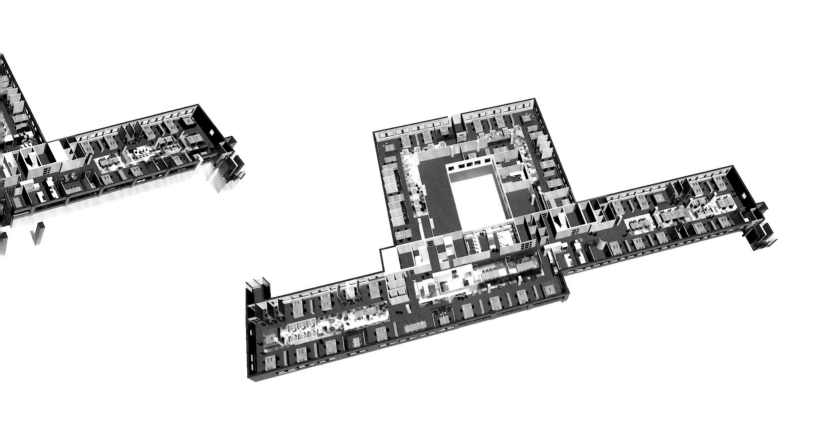

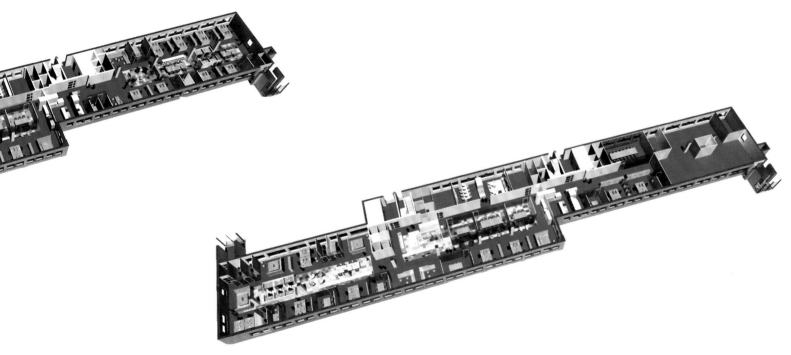

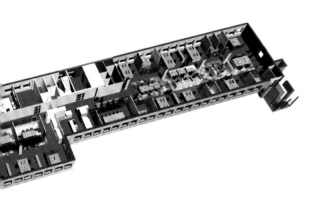

现代企业在不断地发展，它们的核心理念就是灵活，即富有适应性和柔韧性。这些词带有隐喻性，它可以用来比喻一个人的性格，比喻公司或建筑，这也是时代发展的要求。

交流工具的不断发展使人们可以在任何地方工作，固定的办公空间不再适合这个时代。新的工作方式的出现要求有不同的工作环境。例如，只考虑在活动的基础上设计工作区，集中在一个安静的房间里工作，在一个鼓舞人心的房间里进行头脑风暴，磋商室就是一个会议室，甚至在家工作也可以包含在这个概念中。灵活地布置工作场所给公司带来了额外的好处，那就是有助于公司更有效地利用空间。

要把这个概念运用到实际工作中，不使

布局过于死板僵化，室内设计的新功能的实现需要多方面的支持。绿色写字楼、可持续建筑，从摇篮到摇篮的概念，这正是我们感兴趣的。

越来越多的人在关注这些主题，他们将项目的构架定义为耐久性。可持续建筑的基础就是设计一个良好的生活和工作环境，并且低能耗，由此方能减少对有限资源的使用。

这些设计理念除了对自然环境和周围环境起到积极作用之外，特别是室内空间的设计理念对材料的选择，对办公室内环境也有积极影响。研究表明，改善室内环境能提高员工5%~10% 的劳动生产率。

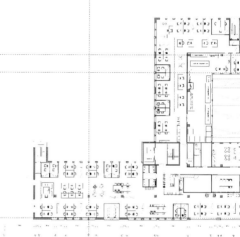

Second Floor

First Floor P

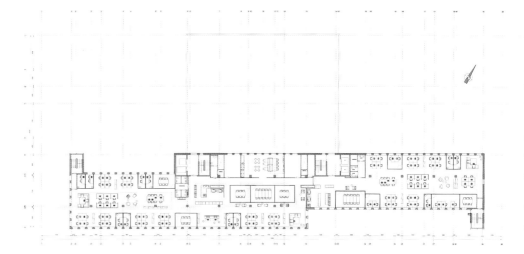

Fifth Floor Plan / 五层平面图

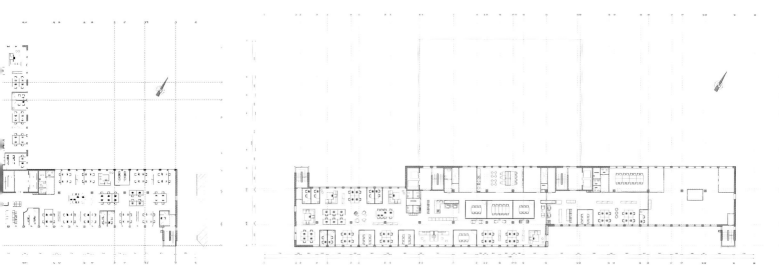

Fourth Floor Plan / 四层平面图

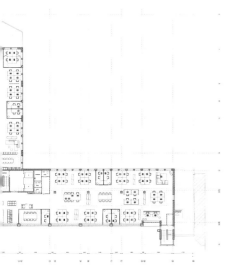

Third Floor Plan / 三层平面图

Pplusp Designers Office
维斯林室内建筑设计有限公司办公室

Design Company: Pplusp Designers, Ltd.
Designer: Wesley Liu

设计公司：维斯林室内建筑设计有限公司
设计师：廖奕权

Different from traditional offices of cubes and white light, this office is designed with a home environment and England style.

At the entrance of the meeting room lies a platform, below which there is a Japanese style shoebox, and as at home, staff will change into slippers here and walk into the meeting room which is filled with mild light and light music. In the meeting room, pot plants and gentle aroma relax the working atmosphere and motivate inspiration. At the corner of the corridor, on the wall hangs a white glass drop-shaped blown pendant lamp from Murano islands. GLO Incasso, a pendant adds interest to the space. The flexibility in the use of mirrors and glass is rather creativity and improves the sense of space on the corridor.

The working space is mainly in light colors. The workbench in light color and the projector set off the ceiling in light bluish-green. Near the toilet there is a conspicuous red-brick wall. Traditional workbenches are replaced by long benches, which get designers closer to each other and encourage communication. Behind the workbenches, ping-pang tables will relax and cheer up the staff.

Large windows make the best use of daylight. On the red-brick wall, you can see London railway signs. Wash basins are transformed from vases and are painted by the designer. Ground spotlights, lighted round mirrors, and a big metal mirror circled by vines add vitality and art to the space.

The designers' private rooms are behind the ping-pang zone. There are also large windows welcoming daylight. Tea pot-shaped pendant lamps, old-style wash basins, jeans-covered leisure chairs, antique decorations and furniture, and a wall made from cork wood demonstrate a rich layer of space.

An environmental protection concept can be seen in the power saving lighting design, large windows and space division.

LED lights which are efficient and energy-saving are used to replace traditional halogen lamps which are harmful to the environment.

The smart plan of space division avoids energy waste from conditioners. The space is divided to a meeting room, offices, a library, a ping-pang zone and toilets. This clear division avoids power waste from the overuse of conditioner so that the AC is only on for those spaces in need.

设计师深信一个有创意的工作环境能够启发思维和鼓励创新，所以工作室舍弃了传统的办公室设计，去除了四四方方像城墙般的间隔与以白色灯光作为主调的空间氛围，而采用以家居氛围和带有英伦风格意味的元素去设计工作室。

为达到家的舒适感觉，设计师在会议室走廊处设计了一个升起的地台，地台下是日式的鞋柜。进入工作室，所有客人和同事都可以在这里伴着柔和的灯光和轻音乐换上拖鞋，像回家一般温馨。小盆栽植物的点缀既为这个密闭的空间提供新鲜氧气，与又清淡的花香气互相呼应，营造激发创作力的空间气氛。走廊转角处悬挂着的是在威尼斯穆拉诺岛手吹制作的白色玻璃水滴吊灯，为空间提供充足的照明。GLoIncasso 既是一件艺术品，也是放在悬浮的"水滴"下面的形态自由的玻璃片，增强了空间趣味性。为了增加空间感，镜子与玻璃的灵活运用是走廊的另一个亮点，镜子背后的组柜，提供了更多的收纳空间。

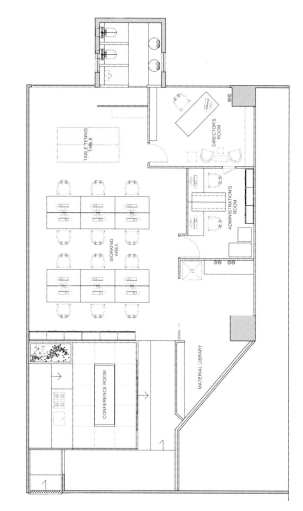

Site Plan / 平面图

工作区以浅色系为主色调，整齐排列的浅色工作台、射灯与浅蓝色的天花相映衬，让人有蓝天白云的自然舒适之感。靠近洗手间的墙壁采用了一般在户外才用的红砖，十分耀眼。此外，长形工作台取代了传统的工作台，拉近了所有设计师之间的距离，鼓励着设计师间的交流与沟通。为了刺激设计师的创作灵感，在长形工作台后面设置了乒乓球桌，供员工们休闲娱乐，抖擞精神。

设置在工作区尽头的大幅玻璃窗，使空间自然采光极佳，避免了白天不必要的照明。从走廊到洗手间一路走过，便会发现钉在红砖墙上的伦敦火车铁路牌，以及设计师亲手刷上的油漆以花瓶改造而成的洗手盘，更有小小的坐地射灯和周边亮着光的圆镜子与相配合。被树藤缠着的另一块金属大镜子仿佛处于花园之中，为空间加添了生命力和艺术感，让设计师在有限的空间下享受最自由舒适的创作环境。

乒乓球桌后面是设计师的独立办公室，大窗户的设置让日光更好地进入室内。有趣的茶壶吊灯、旧式的洗手盘、牛仔布休闲椅和其他古玩的小装饰和家具，呼应着软木制作的墙体，产生别致的视觉效果。不同物料的运用，使空间展现出了丰富的层次感。

该工作室的设计环保，大窗户的使用极大地降低了能耗，空间分区的方式也在不同的方面节省能源。

工作空间的照明设计以不损害环境为前提。为了取代传统的卤素灯，设计师采用了节能高效的LED反光灯。这种对环境无害的照明设计不仅有助于节省电费，而且避免了卤素灯产生热量的问题。

此外，空间明确划分成不同功能区域，如会议室、工作空间、图书馆、乒乓球区和洗手间。有了这种明确的空间分区，空调可以只在活动的地区开启，这有助于避免因过度使用空调而造成的电力浪费。

Italia Telecom Office
意大利电信公司

Design Company: Studio Andrea Castrignano
Designer: Andrea Castrignano
Photographer: Matteo Cirenei
Area: 1,000 m²

设计公司：Studio Andrea Castrignano
设计师：Andrea Castrignano
摄影师：Matteo Cirenei
面积：1 000 m²

If you think that redesigning an office could be less complex than an house you're wrong. This project, for its importance and for its standing, is the evidence. For me it has been a challenge in a challenge. Most of all because I wanted to use art as a vehicle for rethinking the function of this place: from a working venue to an "exhibition" full of precious elements of historical importance. All has been possible thanks to the archive of artworks (50 available to us) part of the Dr. Olivetti's collection. From personal collection to heritage on display, as art lover and later as gallerist, from a museum to an office...I tried to mix the parts and reassign the rules: an office that is also an accessible archive able to attract those who visit or work there, a collection of modern art to live and observe day after day. Rooms are dressed in white to leave space to works' colors that become decorative elements in a space that doesn't want to be aseptic but totally dedicated to what is home.

如果你认为重新设计一个办公室比重新设计一个家居空间简单得多，那你就错了。对设计师来说，这是一个极具挑战性的项目。在设计中，设计师以艺术作为一种媒介，重新思考每个空间的功能，从办公区域到充满历史感的重要"展览品"——Olivetti博士收藏的部分珍贵收藏。空间中有的是个人收藏品还是展出的文物？是博物馆还是办公室？设计师试图对空间中各种元素进行整合与规划，使其具备现代艺术的特征，并经得起时间的推敲与打磨，打造出一个吸人眼球的精彩空间。设计将空间都刷成白色，让生活在其中的人们为其增添色彩，使其自成空间装饰元素，希望它给人的并非冷漠的感觉，而是让人们体验到什么是家的感觉。

Red Bull Office
红牛办公室

Design Company: pS Arkitektur
Designer: Peter Sahlin, Therese Svalling
Participant: Oliver Soderlund, Lina Mezquida Forsnacke
Lighting Designer: Beata Denton

设计公司：pS 建筑公司
设计师：Peter Sahlin、Therese Svalling
参与设计：Oliver Soderlund、Lina Mezquida Forsnacke
灯光设计：Beata Denton

The new Red Bull office in Stockholm is designed with an elegant palette of blue, black, gold and natural materials such as marble, leather and wood. It has taken its inspiration from the slim Red Bull can and the company's enthusiasm for adventure and art. The need of flexibility for an office that likes to invite artists and guests to their workplace has made the design at parts movable, foldable and playful.

Red Bull with its sun, rhomb, bulls and font has given inspiration to the design. Leather furniture (bull), round furniture (sun), rhomb pattern in carpet and furniture and thin black lines (font) recurs in the office such as frames and furniture legs. The reception and lounge is designed with an emphasis on flexibility. The custom designed reception counter is foldable and movable. You can place it and fold it to fit different purposes. Most of the furniture here is easy to move around. You can spin around and rock from side to side in the round sculptured seats or swing in the circle swigs.

Red Bulls enthusiasm for sports, adventure, music and art among other things is shown in installations in meeting rooms where pictures of their opinion leaders is shown.

斯德哥尔摩的红牛新办公室由优雅的浅蓝色、魅惑的黑色、高贵的金色组成。所有的材料都是天然材料，如大理石、皮革、木材等。其设计灵感来源于废弃的红牛罐，设计理念与红牛公司的冒险精神及对艺术的热情相符合。由于办公空间常常有艺术家和客人来访，故灵活性与艺术性就显得尤为重要了，也要求设计具有可移动性、折叠性以及趣味性。

空间内家具的设计灵感来自于太阳、菱形、公牛和圣洗池，如办公室里的皮革家具（公牛）、圆形家具（太阳）、菱形的地毯、办公桌的桌腿和样式等，都得益于此。前台和休息室的设计主要彰显空间的灵活性。前台的柜台是可折叠、可移动式的，既可以把它折叠放置，也适用于其他各种用途。这里的大部分家具都是可移动的，你可以坐在圆边的雕刻椅子上从这头滑动到那头，或在摇动的椅子里痛饮。

会议室里挂着公司领导的照片，彰显着红牛团队对体育、冒险、音乐和艺术的热衷。

Optimedia Media Agency Office
Optimedia 媒体机构办公室

Design Company: Nefa Architects
Chief Architect: Dmitry Ovcharov
Authors Team: Dmitry Ovcharov, Maria Yasko
Architects: Victor Kolupaev, Olga Ivleva
Photographer: Ilya Ivanov
Area: 500 m²

设计公司：Nefa 建筑公司
主案建筑师：Dmitry Ovcharov
设计团队：Dmitry Ovcharov、Maria Yasko
建筑师：Victor Kolupaev、Olga Ivleva
摄影师：Iya Ivanov
面积：500 m²

From the architects. "Our studio Nefa Architects has had plenty of experience in designing working spaces for ad agencies and media companies."

"This time, developing the concept for Publicis Group, that includes 8 companies, we have opted for an idea of office spaces as modern art galleries. We did not want to make something complex; in each project the compositional idea, serving as a purely artistic feature and exposition, particularly, within the context of a brand book, organizes the space".

"When designing the working space for Optimedia media agency office, we used saturated colour, central to the company's signature style, and optical effects, 'breaking' the real space and creating a parallel reality."

"The interior may or may not get some additional drawings or slogans or a new color in the future. That is the idea. The environment may change, but what will always be maintained is the core expressive substance, which is self-sufficient and makes the space complete."

从建筑师队伍来看，"我们工作室的建筑师空间设计经验丰富，他们能为广告公司和媒体公司提供优质的服务。"

"这一次，有8家公司为 Optimedia 媒体机构做概念设计。而我们倾向于现代艺术画廊式的办公空间理念，我们简化空间设计，以一个品牌的背景为中心，组织空间的设计，释放创作灵感，把每一个设计都视为纯粹的艺术特色创作。"

"在为 Optimedia 媒体机构设计办公空间时，我们使用饱和色，达到了该公司鲜明的视觉冲击的核心需求，打破了原有空间局限，创造了一个平行的现实。"

"在我们国家甚至是全球有可能使这一概念得以完善，又或许不可能，或在将来也得不到认可和发展，但这就是我们的设计理念。环境也许会改变，但始终不变的是将核心内容淋漓尽致地体现出来，以设计完善空间。"

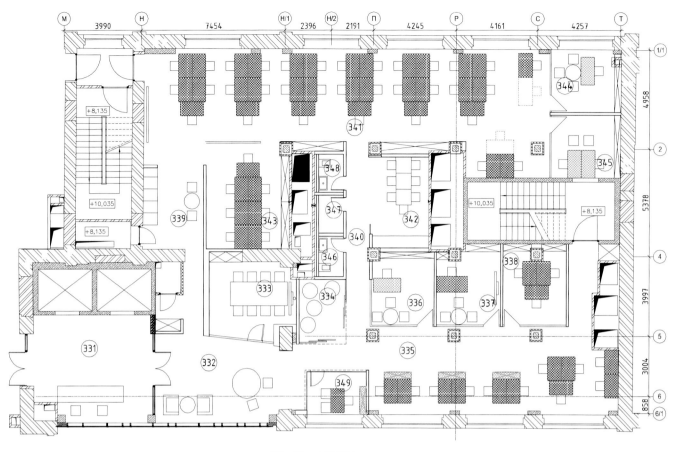

Site Plan / 平面图

Dious Furniture Office
迪欧家具办公室

Design Company: ADD+ Creative Organization
Designer: Guo Weicheng, Zhan Qijun, Liang Yingyi
Area: 700 m²

设计公司：ADD+ 创意机构
设计师：郭为成、詹绮君、梁颖怡
面积：700 m²

What we provide are not only products, but humanistic thinking on office behaviors, care for office workers, and pursuit of perfect office environment.

Hope, penetration, autumn, water are the key words of the project, express design concept.

Hope: It is hope that gives us never-ending power of life. We hope to build a great enterprise which is the goal of us all.

Penetration: We manage to achieve the penetration of light and sight in harmonious spaces. To go beyond ourselves – that is must spirit innovation of modern enterprise. We gaze our projects with eager expectation.

Autumn: autumn is the harvest season. If we do our best, we will get the joy of success.

Water: water is the representative of spiritual and the source of wealth.

The designer are thinking about the exhibition type of office furniture. They start their study from office behavior, to office management, humanization of work procedure, the alternation of working alone and working together. In here, it can meet all the demands by the design of furniture.

Introduce the air, sunshine, water, light and shadow into the interior, and to make the dialogue between the nature and people in the office available. The office environment and our life are inseparable, so a cultural pursuit of life will give us the passion for work. In the same way, A good working environment will improve the value of our products.

Site Plan / 平面图

在本案设计中呈现的不仅是产品，而是对办公行为的人文思考，对办公者的人性关怀，以及对办公环境无限的追求。

设计以"望、穿、秋、水"为关键词，将设计师的设计理念融于整体空间之中。

望——希望。希望给了我们无限生活的动力，希望企业的美好愿景，成为大家共同奋斗的目标。

穿——穿越、突破。空间的交融，光线的穿透，视线穿越。现代企业必备的创新精神，寓意自我突破。

秋——秋天是收获的季节。勤劳的收获及成功的喜悦。

水——水源，灵性的代表，活力与财富的源泉。

设计师一直在思考办公家具的展示模式。从人的办公行为研究开始，到办公管理的模式，以及符合科学的管理，人性化的办公运作流程，员工独与众模式的交替，这些都通过家具的组合方式来实现其功能需求。在这里，展示的不再只是产品，而是对办公这个行为模式深度的探寻。

设计师提倡将空气、阳光、水、光影纳入室内空间，让这些自然的属性与空间的人对话。办公环境与生活密不可分，没有生活的人文追求，哪来办公的激情呢？同样，没有好的引导环境，办公产品如何能提升价值呢？

Victory Furniture Group
百利文仪（中国）家具有限公司

Design Company: ADD+ Creative Organization
Designer: Guo Weicheng
Area: 3,000 m²

设计公司：ADD+ 创意机构
设计师：郭为成
面积：3 000 m²

Experience Lounge of Victory（China）Shanghai branch is located in the Huangpu river wise valley Pioneer Park. The spaciously area is near 3,000 square meters. The design company and the owner made a great effort to prepare this project. It was completed with a year of hard work.
The plan of exhibition room presented the design conception of the designer. The traditional thoughts were replaced by the industrial decoration concepts. The designer offers an integrated solution of business space. The high wall cabinet shows a new milestone of Victory Group.

百利文仪（中国）家具有限公司上海体验厅坐落于花园式的浦江智谷创业园，面积近3 000平方米，空间敞朗弘阔。经过设计公司与业主多方的努力与配合，用了将近一年的时间筹备与组建而正式落成。

从展厅的规划到设计，设计师都以工业化装潢替代传统战略思维，极力呈现"商务空间的整体解决方案"。展厅内高隔墙柜的应用设计，为百利集团活态商务空间的落地，树立了崭新的里程碑。

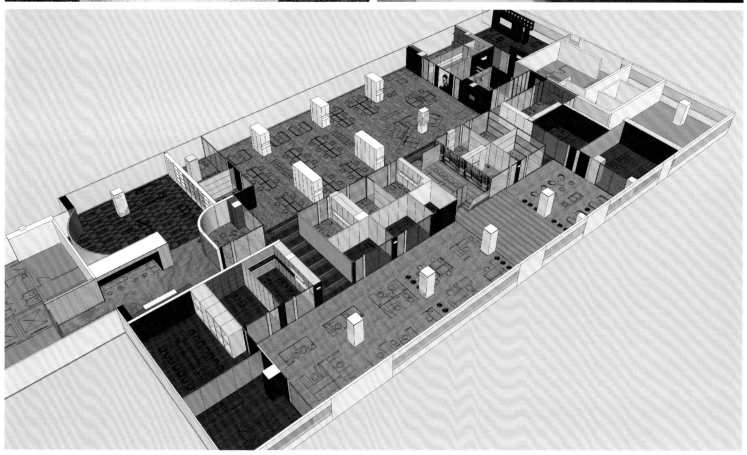

Zhongliang V City Model Office
中梁 V 城市办公样板房

Design Company: Shenzhen Yipai Interior Design Co., Ltd.
Designer: Duan Wenjuan, Zheng Fuming

设计公司：深圳市伊派室内设计有限公司
设计师：段文娟、郑福明

The target clients of this project are listed group companies of high-tech industry. The first floor can be used as a club and the second floor is office area. In this project, there is a reception hall, office areas, meeting rooms, display areas for project models, a drink bar, a manager's office, a general manager's office and an accounting department.

In the lobby, geometric patterns on the ceiling and walls construct an environment of science and technology. Next to the lobby stands an exhibition area of company products and culture, and a projector here enriches the area. The drink bar provides business services for product launch events, activities, business talk and reception and it is also a place for staff dining, relaxation and parties. On the second floor, geometric patterns on the pillars, ceiling and walls echoes that on the first floor. The golf ball and private collections in the chairman's office demonstrate the quality taste of the chairman. The design styles of the wall and the ceiling match well with each other so that the space is in a harmonious environment.

本案是中梁 V 城市的办公样板间，定位于高科技产业相关的上市集团公司。由于公司常有会客，一楼兼具会所功能，二楼办公。根据公司的办公性质和办公需求设置了接待大厅、办公区、会议室、项目模型展示区、水吧、经理室、总经理室、财务部等。

由于大堂是中空区，视野开阔明亮，顶面与墙面的几何图案营造出一种具有强烈科技感的空间氛围，大堂区紧邻着产品及公司文化展示区，投影灯的展示丰富了展区内容。吧台区既有用餐、休息、洽谈、接待等功能，也可作为产品发布会、聚会的场所，实现商务生活一体化。二楼的柱子以及天花、墙面的几何造型延续一楼的，空间整体风格统一协调美观，营造简洁、大胆的办公场所。董事长办公室的高尔夫球杆、收藏品等不仅体现出个人喜好及收藏品位，也体现出不一样的高端生活享受。

Creative Space Office
创域办公室

Design Company: Shenzhen Creative Space Decoration & Design Co., Ltd.
Designer: Yin Yanming
Area: 680 m²
Main Materials: Stones, Black Steel, Pvc Floor, Textile Wall Covering, Coating, Glass, Wooden Veneer

设计公司：深圳创域设计有限公司
设计师：殷艳明
面积：680 m²
主要材料：石材、黑钢、胶地板、墙布、涂料、玻璃、木饰面

We moved to a good place of flowers and birds. It is in the southeast of the building with plentiful daylight and a good view.

Offices should be a place full of emotion, and a mirror of our company.

We lay emphasis on space arrangement and the design light. The orange lamplights on the corridor project the company's logo inlaid in the dull-polished glass doors on the marble ground. The workbench and background wall of the reception area are made from finger-jointed wooden veneer. On the other side, a pair of lintels of late Qing Dynasty are hanging on the wall. You can see moiré and floral scrolls on the ebonized lintels, below which stand a pair of Chinese style arm chairs of high backs. On the acrylic tea table, flowers in bright yellow stand gracefully. All of these draw a Chinese painting of space humanity.

A corridor divides the reception and the office area. Two collective paintings face each other at the ends of the corridor. Lotuses and hydrangeas illuminated by gentle light bring about a humane odor and eastern romantic charm. Several hollowed-out white screens separate the reception and offices naturally. White workbenches are arranged in orders. The ceiling with simple and clean design is of structural beauty, and a chandelier enriches the space layer. Plentiful sunshine and gentle light ease the toughness and coldness of large-scale whiteness in the space. Glass separates the manager's office from other spaces. The one-way-vision glass decoration connects the interior and the outside space. The spacious material room and operation tables in it make it available for the designers to demonstrate their design dreams.

The office is of 50 square meters. Facing the south, the teakwood office table is in simple design. Behind the table stand a large-scale hand-sketching screen on which there is an abstract painting of a foreign artist. In the front of the office table, custom-made sofas in beige and black circle welcome the guests to have a rest. A white floor lamp gives light to the white butterfly orchids on the tea table. There is a tea area in the office. The tea table is in simple shape and smooth lines.

Site Plan / 平面图

The terrace is the favorite of my staff and me. On the main walls sketch the work of children interest and imagination, which create lively and leisurely scenes of life. On one side of the terrace, there is a wooden floor, a bamboo groove, garden tables and chairs, and some flowers and plants, which is a good place for leisure.

The design materials are fairly common, but the designers' experience, cognition and emotion endow the materials with life and thus build a special space quality. The beauty of design sprouts from heart and we will continue to perceive our dreams here.

机缘巧合，多方寻觅，终于在一个风和日丽的日子里从六楼搬到了五楼，从大厦的西北角来到了光线充足、视野开阔的东南向。虽然没有"面朝大海、春暖花开"的惬意，却也拥有了80平方米露台的花香鸟语，也算了却了这几年的一个小心愿——能在工作之余有一处悠然自得的好去处。

这是自公司成立以来第三次装修自己的办公室。每次都是自己动笔，每次却也都是经历了提笔、放下、再提笔反反复复的过程，在纠结中让落在纸面上的点和线生成空间，产生意义的时候，不觉就想起一句唐诗来"谁家新燕啄春泥"，我大概也是这样的吧，一点一滴衔来树枝、软草和泥土慢慢筑就自己的梦想。这680平方米的空间在图纸上横着竖着看，也就是几个标着尺寸的方方正正的"盒子"，但是想想我们一天会有多少时间在这里度过呢？你就越发觉得这里不仅是工作场所，而是企业的一面镜子，更是充满情感的场域。虽然笔下的硬装、软装，一旦实施，你爱或不爱、你看或不看它就在这里，但是我更愿

意它们所构筑的场域能把那些心底的喜怒哀乐和快意人生都静静地表露出来。

这是一个两户打通的空间，方正、通透。空间功能布局并无多少犹豫，而灯光营造成了设计的重点。大楼走廊橙黄色的灯光透过公司磨砂玻璃门，把门上镶嵌的公司LOGO斜斜投射在入口玄关的大理石地面上，一切的游走便从这里开始。接待区的工作台和背景墙由指接板饰面，深浅相间的棕色木纹提高了空间感知的温度；而相对的另一边则在整墙上横挂着一对晚清门楣构件，乌木色的雕工镂刻着精致的云纹卷草，下摆一对中式高背靠椅，方正的亚克力透明茶几上跃动着簇簇艳黄的虎跳兰，寥寥几笔勾画出的这组中式小景奠定了空间的人文基调。

一条横向的走廊在布局上分隔了接待空间与办公区域，两幅珍藏版的云南画派原作在走廊的两端遥遥相对，点射凝聚的那束光照亮了一朵朵由黄、白、绿濡染出的莲花、绣球花，几分水墨淋漓恍然若梦，又有几分高雅恬静，细致

婉约地演绎出空间中的人文气息与东方神韵。几幅间隔有序的镂空白色喷漆屏风在走廊边把接待区与办公区自然隔离。线与面的组合运用是贯穿中间开敞式办公区域的主要设计手法。白色的工作台在空间中整齐排列，天花造型简单干净显露出结构之美，悬浮的线型吊灯增加了视觉层次。南面排窗带来了充足阳光，结合工作区吊灯细心选择的色温，有效调节了大面积白色带来的视觉硬朗感，营造出了宽敞明亮、简洁而不失温暖的工作氛围。周围玻璃隔断出独立的主管办公室，单孔透的玻璃贴饰让大空间与小空间之间有了另一种虚实呼应的感受。灵活组装系列的家具，让后期的施工周期与细节处理游刃有余。最舒适的是材料室，偌大的空间和操作台面可让设计师细细排列、选择、构筑，展现其设计思想。

办公室设计画龙点睛之处自然是我的办公室，面积五十平方米，舒适开阔，窗外绿草如茵。柚木长台的办公桌坐北朝南，线条简洁、朴实自然。背后大尺度白色娟质手绘屏风上，国外艺术家的写意抽象画遒劲有力。办公台前，定制的米白色和黑色皮质沙发围合出一方休闲区域，一根纤细而有韧性的黑钢条弯曲得恰到好处，斜斜挑出一盏棱角分明的白色落地灯，正正的关照着中间大理石方几上几支伸展的白色蝴蝶兰，营造出柔和雅致的洽谈氛围。随着年龄的增长，在年少轻狂之后大概会有一条回归之路，对传统，对故土，或许这就是一种精神上的依托和安定。我专门在办公室一隅开辟出的一方茶室，放着一台从故乡云南普洱淘来的千斤沉水普文楠茶台，它造型简洁，形体线条自然流畅，台面木纹和树根影纹实属上品。放上各式茶具，养花、焚香、泡茶，和来此小坐的朋友们谈天说地、品茶论道，真正是"有朋自远方来，不亦乐乎"。

走廊折角再拐弯，便是半户外的露台，这里是员工和我的最爱。两面主墙上留下了设计师们充满童趣与想象力的手绘作品，如历险的蓝丁丁、纵情跳跃的斑马、角落里逍遥自在的板凳人和大树下嬉戏的小狗，形成矛盾三维空间的动态，也建构出朝气、闲适的生活场景。露台的另一边，铺陈的木地板与墙边排列得错落有致的竹林相得益彰，再放上几组花园桌椅，几盆花草树木，就都齐了。只要愿意，阳光在这里能走多远就走多远；只要愿意，清风在这里能摇曳徘徊几圈也都随意；只要愿意，员工们可以在这里休息、午餐，喝喝下午茶，三三两两在桌球台边一试身手。完工乔迁适逢新年，我们的设计、成长与梦想就在这里开始，从我们身边的这一方听风、看景的小小天地开始。

因为是自己的办公室，所以从设计构思到施工材料、工艺流程都要细细掂量考虑周全。整个设计选用的材料普通至极，如墙布、涂料、金属、玻璃，但是设计师积累的生命体验、设计认知和发自内心的情感赋予了它们新的生命，构筑出特有的空间气质。设计之美从心里开始，每天都有新感觉，怀着一份好心情让设计继续在这里筑梦，让更多好作品从这里开启，当然，希望幸福不是一个人的，而是在大家的心里，也包括正在阅读的你！

284–319

3 Loft Office Space
LOFT 办公

Zhongliang Model Office C
中梁 C 户型办公样板房

Design Company: Shenzhen Yipai Interior Design Co., Ltd.
Designer: Duan Wenjuan, Zheng Fuming

设计公司：深圳市伊派室内设计有限公司
设计师：段文娟、郑福明

The project is aimed at attract young, fashionable and aspiring SOHO people.
On the first floor, there is an office area and an open kitchen which is also a drink zone. On the second floor stands a private study, which is also the manager's office, a rest room and a leisure area which is also a guest room. Beautiful iron stairs and photo wall divide the office and living areas.
As for decoration, a lot of materials of original ecology are used, such as raw wood, stones, iron artwork and red bricks, to create a special and free space of original ecology. In the office, the back ground wall is decorated by raw wood sections. The wall is decorated by a customized China may of wooden plank and some designer's handwork. A climbing wall and swings beside the windows make an interesting environment. Under the stairs, mountain bikes and a firewood on which the names of cities where the proprietors have been to are carved demonstrate the masters' life and work style. The gramophone on the second floor plays music of different rhythm at different time.

如今SOHO已经成为越来越多的年轻人追求的生活和工作方式，他们推崇自由、浪漫的工作状态，家是SOHO族梦想与生活共存的地方。本案便针对追求时尚与个性、自由与理想的年轻SOHO一族。

一楼是办公区域与开放式厨房（兼水吧区），二楼是个人书房（兼经理室）、休息室与休闲区（兼接待室）。铁艺楼梯、照片墙过渡办公与生活区域。

本案在装饰上大量采用原生态的装饰材料，如原木装饰、石材、铁艺、红砖等，营造出个性、自由、原生态的空间。办公区背景墙采用原木切片作为装饰，体现生态、个性、绿色主题，墙壁采用定制中国地图实木板装饰结合设计师简易手工，将线与城市连接，体现出工作室的业务辐射。窗边的攀岩墙及秋千，除了营造空间氛围外，还极具趣味性，也展现了层高的建筑优势。楼梯下面的山地自行车及刻有主人到过城市名称的柴火堆都是生活及工作方式的体现；二楼的留声机在不同时间播放着不同节奏的音乐，给参观者不同声音的触动。

Chengdu Vanke Diamond Plaza Loft
成都万科钻石广场 LOFT

Design Company: Shenzhen Creative Space Decoration & Design Co., Ltd.
Designer: Yin Yanming
Area: 120 m²

设计公司：深圳创域设计有限公司
设计师：殷艳明
面积：120 m²

The design inspiration of this project stems from the designer's yearning for nature, thus irregular shapes and changes of blocks and surfaces are used to interpret the beauty of nature.
There is a rhythm beauty in the office where spaces changes are made by design techniques of zigzag, slant and cutting. Between the entrance and the antechamber, a slant cutting conveys vitality and movement to the space. Space tones of the antechamber and working areas are divided by a bold cutting. The atrium part makes the office special. On the second floor, the boss's office and the meeting room which is also a creative space are not separated for the convenience of communication. Black, white and grey set the main tone, while green infuses a fresh breath to the space, which vivifies the atmosphere and echoes the design concept.

本案的设计灵感来源于对自然的向往与追求，自然万物各有其形态与节奏。自然是一种无形、抽象的事物，所以设计师在本案中利用不规则形体与块面的变化来诠释大自然之美。

原有室内空间相对方正，设计师以建筑空间手法入手，通过曲折、斜向及切割体现关系的变化，使空间产生一种节奏感和韵律感。从入口至前厅，一个斜向的切割打破了原有空间的沉重感，使得空间更具活力与动感。前厅与办公区采用大切割手法打破整齐划一的空间色调，结构之美无处不在。中空的挑高作为垂直空间的移动动线和视觉上的整合过渡空间，干净利落地打造出一种新颖、独特的 LOFT 空间。二楼总经理办公室与会议室（创意空间）不分区，便于整个团队成员之间的沟通，有助于增强团队的凝聚力。企业以电子制作、广告设计等业务为主，故空间以黑白灰为主调，使空间有颜色深浅对比，强调时尚感、空间感与设计感。局部点缀一些跳跃的色彩（如绿色），为空间注入休闲的气息，活跃了整个空间，又实现了工作与生活之间的健康平衡，呼应了设计主题。

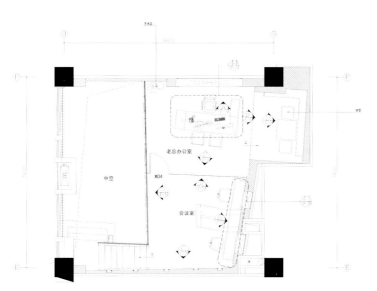

Interlayer Plan / 夹层平面图　　　　　　First Floor Plan / 一层平面图

Pearl River Technology Digital City LOFT Apartment
珠江科技数码城 LOFT 公寓

Design Company: C&C Design Co., Ltd.
Designer: Peng Zheng
Participant: Wu Jia
Area: 124 m²
Main Materials: Bricks, Wood Veneer, Stoving Varnish, Carpet, Stainless Steel, Glass

设计公司：广州共生形态设计集团
设计师：彭征
参与设计：吴嘉
面积：124 m²
主要材料：砖、木饰面、烤漆、地毯、不锈钢、玻璃

How to build an all-round office in an apartment of less than 70 square meters? The designers prove the possibility of building a multifunctional space having a capacity of business and residence. Being located at the northeast of Foshan and adjacent to Guangzhou, the project is surrounded by small and medium-sized enterprises. In this small auto parts company, there are areas for display, discussion, offices and toilets on the ground floor. A meeting room and a manager's office are on the mezzanine. After transform, the apartment is enlarged to 124 square meters.

White set the main tone of the space, and the carpet in dark grey intensifies contrast of space brightness. Simple colors and flat designs construct a pure and clean environment of an international working space, and the atrium makes the office well lighted and ventilated.

如何将一个不到 70 平方米的公寓设计成一个功能齐备的办公空间？设计运用 LOFT 的设计手法解决功能要求，同时为目标客户展示了一个多功能的商住空间。项目位于广东佛山东北部，毗邻广州，周边多为中小民企，如服装、汽配、轻工产品及材料配件行业等。本案为一家小型汽配代理（研发）公司，一楼包含展示区、公共办公区、洽谈区和卫生间，夹层则包含会议室和经理室，改造后的公寓面积由 67 平方米扩大至 124 平方米。

设计以现代简约的白色为主色调，深灰色的地毯强化了空间的明度对比和张力，高调简洁的色调和扁平化的设计风格营造出一种纯粹、干净的国际化办公空间氛围，挑空的中空让空间形式更加丰富的同时也便于采光和通风。

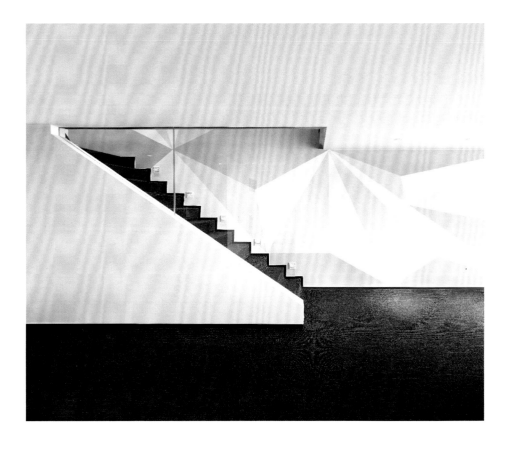

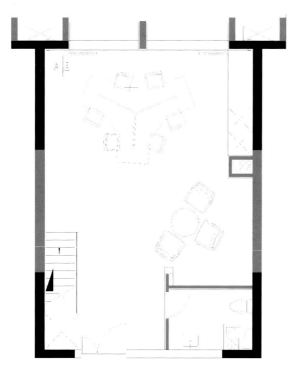

First Floor Plan / 一层平面图

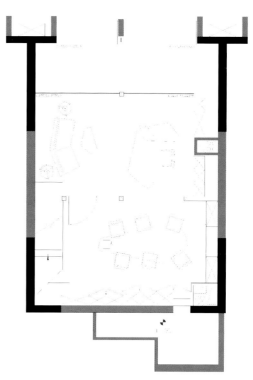

Second Floor Plan / 二层平面图

Tianjin Meinian Square Loft Model Office

天津美年广场 LOFT 办公样板间

Design Company: Shenzhen Creative Space Design & Decoration Co., Ltd.
Designer: Yin Yanming, Wan Pan
Area: 260 m²
Main Materials: PVC Floor, Wood Veneer, PVC Woven Carpet, Leather, Stainless Steel

设计公司：深圳创域设计有限公司
设计师：殷艳明、万攀
面积：260 m²
主要材料：地胶板、木饰面、PVC 编织地毯、皮革、不锈钢

This is a clothing trade company. In order to highlight the good integration of work and life in a rather open and interesting office space, the designers draw the conception of "folding planes and cross lines" into the design of the office.

Pillars are used as space partitions. To create a large-scale dynamic loft space, the designers spent much thought on the partition of functional spaces. A ceiling-tall background wall at the entrance of the office promotes the spatial scale of the loft space. Shapes of folding planes both on the ceiling and at the reception create an atmosphere of fashion and vivacity. The whole space is extended with the use of a grey mirror. In the open working section, apart from areas of common work, there is a discussion area and "make it fun" working area, which brings about openness and flexibility to the office.

Interlaced ceiling grooves and a chandelier make the space vivid and interesting. In the chairman's office, zigzag forms drive out stiffness and depression of upper floor. Shapes of folding planes here echoes to the design of the hall ceiling. Under the designers' hands, the pipeline equipment floor above the aisle and the space below the stairs become zones for showcase and model display. This maximum use of space demonstrates enterprise culture. The design conception of "folding planes and cross lines" is applied with different techniques in different spaces of the office.

Black, white and grey set the main tone of the space, while embellishments in yellow and green bring about air of relaxation and coziness. The shape design of furniture also exhibits the idea of "folding planes", which adds flexibility and variation to the office.

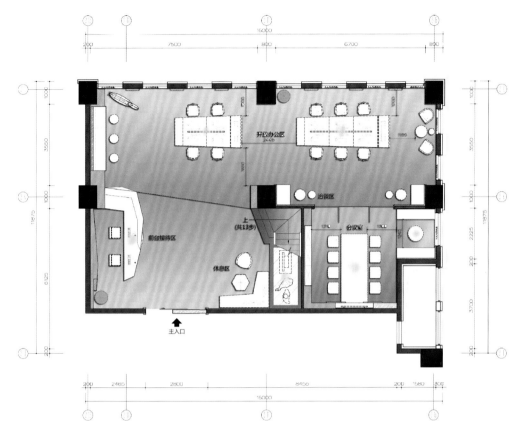

First Floor Plan / 一层平面图

现代办公空间设计着眼于潮流发展趋势、工作观念的改变等方面，体现工作与生活的有机融合，以及空间的开放性和趣味性，与其他方正、中规中矩的空间有着天壤之别。因此，设计师在设计的构思中引入了"折面与交叉线形"的手法来打破传统的思维模式。

在原建筑的结构上框架略显方正，整个布局被柱子划分较散，同时作为LOFT空间布局净空会显得低矮。如何打破常规打造一个大尺度的LOFT动态空间，设计师在功能空间的分隔上做了深入的分析和探索。在入口运用挑空处理，直接延伸至天花的背景墙提升了整个LOFT的空间尺度感。天花通过折叠面的造型，为前厅及前台营造时尚灵动的空间氛围。接待台的连体折叠切割面造型不仅时尚简洁，更体现了线面结合的与众不同与现代时尚感。衔接夹层的灰镜折面，延展了空间在宽度与高度上的视觉感受。设计后的开放办公区在原有普通办公的功能上增加了趣味办公和洽谈功能，也增强了空间的半开放性和灵活性。天花线性凹槽的交错分块与吊灯高低错落叠加，化解了低矮空间的压抑感，使原有单调的空间变得灵动。董事长办公室中空间划分灵活通透，打破了传统办公空间布局。局部折形上反式天花的形体，不仅能烘托氛围和解决层高压抑的问题，更多形体块面的折叠与大厅天

花的元素彼此相呼应。部分无法使用的空间，其设计也别具一格，如走道上空管道设备层和楼梯下方空间都设计成了橱窗、模特展示区，既使空间利用最大化，又展示了企业文化与企业形象。空间中所有形体的折面和交叉线性都是通过不同的手法互相转换，整体空间的设计语言与视觉表达相辅相成。

整个空间的色调以黑色、白色、灰色为主，局部以黄色、绿色进行点缀，为整个空间注入休闲惬意的气氛。家具局部造型设计也是采用斜面切割和体块搭接的方式，使整个折面交叉的空间变得灵活多变，充满节奏和律动感。

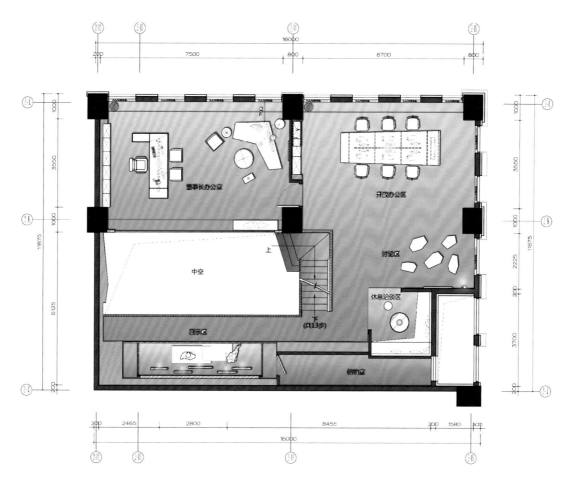

Second Floor Plan / 二层平面图

Regent Taipei Model House C
台北晶华样板房 C 户型

Design Company: Symmetry Design
Designer: Lin Xianzheng
Soft Decoration: MoGA Decoration Design
Area: 105 m²
Main Materials: Wood Floor, White Stoving Varnish, Ultra-Clear Tempered Glass, Volakas Marble, Carpet

设计公司：大匀国际空间设计
设计师：林宪政
软装设计：上海太舍馆贸易有限公司
面积：105 m²
主要材料：实木地板、白色烤漆、超白钢化玻璃、爵士白大理石、地毯

This is an ordered-planned office with huge windows which provide a good chance to enjoy the outdoor scenery.
The multi-layered space links up different working zones. A stair in simplified design, except for its practical function, is also a beauty in this unadorned and elegant space. Wood collages add coziness to the space, and the match of black metal materials and wood veneer finish, whose texture are in contrast with each other, manifests an artistic quality of the space.

本案中设计师有序地规划整个办公空间，错落之中尽显秩序之美。大量玻璃的运用使室内外景观互相渗透，让员工在紧张繁忙的工作时分，也可一窥窗外美景。

而打破桎梏的上下结构，有效地利用了层高优势，让整个空间不同区域间达成沟通对话。楼梯除了基本的实用功能之外，也成了空间里的一道优美风景。简化到极致的楼梯与素雅干净的空间相映生辉。

温馨的木质拼贴工艺更为整个办公空间增添暖意，而黑色的金属材质与木饰面的搭配，以材质上的冷暖对比彰显空间气韵。

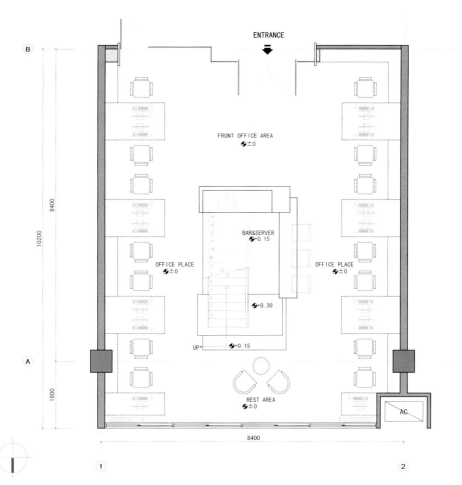

Site Plan / 平面图

LOFT Office
创意LOFT里"V商店"

Design Company: Shenzhen Yipai Interior Design Co., Ltd.
Designer: Duan Wenjuan, Zheng Fuming

设计公司：深圳市伊派室内设计有限公司
设计师：段文娟、郑福明

Different from traditional offices, this office space, infused with cooperation features of electron industry and new media industry, is one of business appeal. This is a bright, spacious and passionate office. Window display at the entrance of the office demonstrates business element of this working space.

Green is the key tone of the office, which represents modernity, technology and creativity. Beside the wall, a tree-shape bookshelf and sports equipments demonstrate a combination of team work and life. Below the stairs stand bar tables of animal shapes. On the wall, heart-shaped team photos build a natural, harmonious and friendly working environment. Decorative pictures on the stairs are pierced logos of different companies which show not only customer resources but also business position.

本案不同于传统的办公样板房设计，设计师结合了电子行业与互联网新媒体产业相关公司的特点，打造出一个具有商业韵味、充满激情的办公空间。整体户型高大而开敞，流动性、透明性都非常好。设计师将户型的特点与现代LOFT风格完美结合。既把LOFT风格代表着的前卫、时尚、创意、自由的生活方式的深层意义传递出来了，又突出了办公商业化。设计师在入门处设计橱窗，展示产品及公司形象。

办公空间主要以绿色为主，代表现代、科技及创新思想。办公区墙面树形书架与运动器材的展示，让工作与团队生活相结合，引领着新的工作及办公模式。楼梯下放置富有特色的动物造型吧台，为空间增添新意的同时，又是别样的办公享受。墙上陈列着的心形团队照片，营造出自然、和谐、亲切的人文工作环境。楼梯上的软装装饰画是由不同公司的LOGO拼合而成的，展示出客户资源及企业定位。设计师的创意点缀着每一个空间，去细细钻研、品味，更能发现每个空间设计中的特殊含义。

图书在版编目（CIP）数据

顶级办公 IV / DAM 工作室 主编 . – 武汉 : 华中科技大学出版社 , 2016.4
ISBN 978-7-5680-1545-5

Ⅰ . ①顶… Ⅱ . ① D… Ⅲ . ①办公室 – 室内装饰设计 – 图集 Ⅳ . ① TU238-64

中国版本图书馆 CIP 数据核字（2016）第 004733 号

顶级办公 IV
Dingji Bangong IV

DAM 工作室 主编

出版发行：华中科技大学出版社（中国·武汉）	
地　　址：武汉市武昌珞喻路 1037 号（邮编：430074）	
出 版 人：阮海洪	
责任编辑：岑千秀	责任监印：张贵君
责任校对：熊纯	装帧设计：筑美空间
印　　刷：中华商务联合印刷（广东）有限公司	
开　　本：942 mm × 1264 mm　1/16	
印　　张：20	
字　　数：160 千字	
版　　次：2016 年 4 月第 1 版 第 1 次印刷	
定　　价：328.00 元（USD 65.99）	

投稿热线：(020) 36218949　duanyy@hustp.com
本书若有印装质量问题，请向出版社营销中心调换
全国免费服务热线：400-6679-118 竭诚为您服务
版权所有　侵权必究